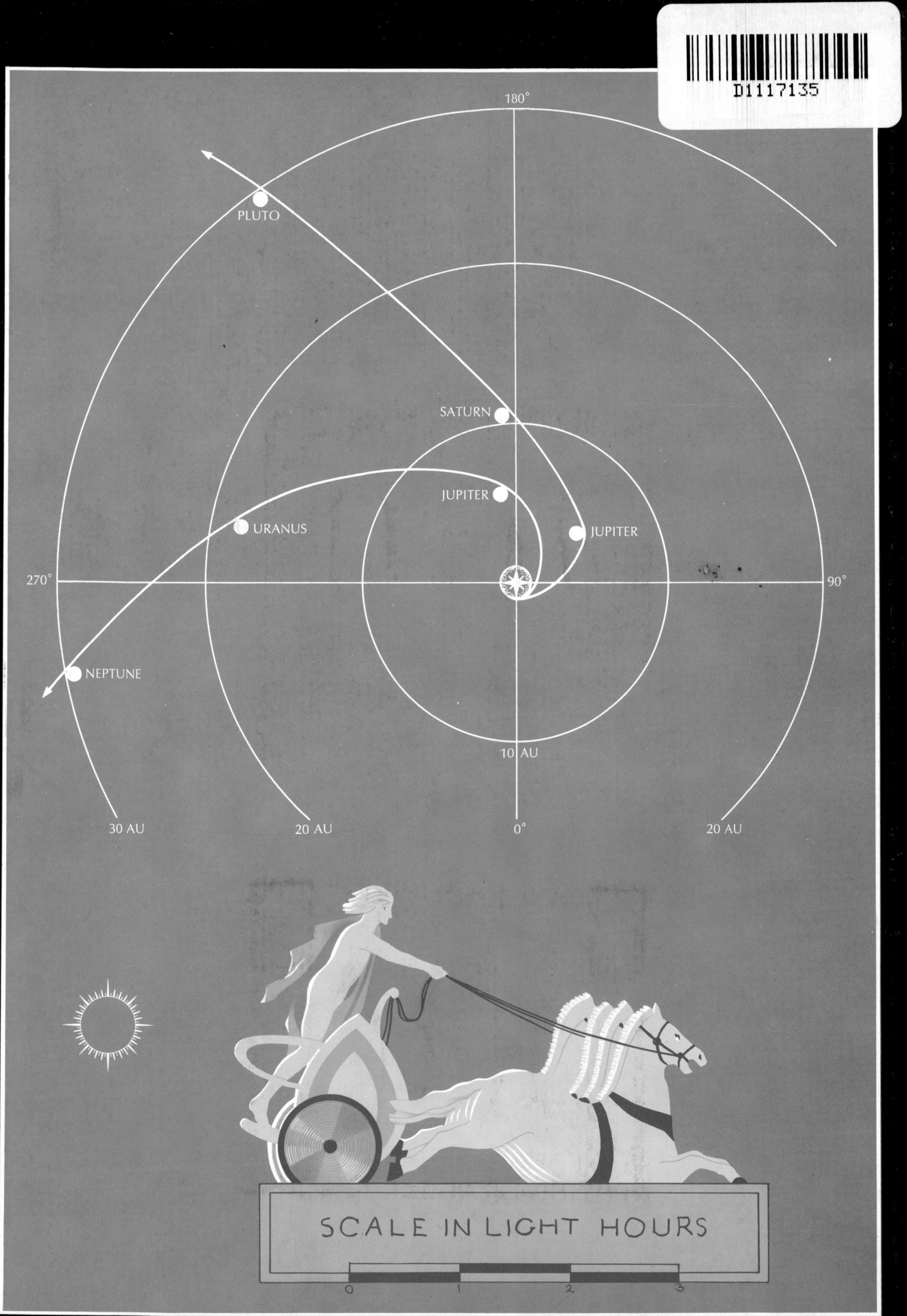
D1117135
180°
PLUTO
SATURN
JUPITER
JUPITER
URANUS
270°
90°
NEPTUNE
10 AU
30 AU
20 AU
0°
20 AU
SCALE IN LIGHT HOURS
0
1
2
3

DATE DUE

OCT 8 '78	APR 21 '82		
DEC 8 '78	FEB. 19 1986		
JAN 26 '77	MAR. 12 1986		
MAR 2	APR. 15 1986		
MAR 23 '77			
JUN 2 '77			
JUL 25 '77			
FEB 27 '78			
MAR 6 '78			
MAR 20 '78			
APR 6 '78			

BOOKS ILLUSTRATED WITH PAINTINGS BY CHESLEY BONESTELL

CONQUEST OF SPACE
by Willy Ley

THE WORLD WE LIVE IN
by the Editors of Time-Life Books

THE EXPLORATION OF MARS
by Willy Ley and Wernher von Braun

BEYOND THE SOLAR SYSTEM
by Willy Ley

MARS
by Robert Richardson

BOOKS BY ARTHUR C. CLARKE

Fiction

THE CITY AND THE STARS

2001: A SPACE ODYSSEY

CHILDHOOD'S END

THE WIND FROM THE SUN

Nonfiction

Epilogue to
FIRST ON THE MOON
by Armstrong, Collins and Aldrin

THE PROMISE OF SPACE

THE TREASURE OF THE GREAT REEF

REPORT ON PLANET THREE

BEYOND JUPITER

BEYOND

THE WORLDS OF TOMORROW

PAINTINGS BY CHESLEY BONESTELL
TEXT BY ARTHUR C. CLARKE

LITTLE, BROWN AND COMPANY—BOSTON-TORONTO

JUPITER

T 02/73

Library of Congress Cataloging in Publication Data

Bonestell, Chesley.
Beyond Jupiter.

Includes bibliographical references.
1. Planets--Exploration. 2. Planets--Pictorial works. I. Clarke, Arthur Charles, 1917- II. Title.
QB603.B65 629.43'54 72-6440
ISBN 0-316-14699-4

First published in 1972 by Little, Brown and Company, Inc.

Published simultaneously in Canada by Little, Brown & Company (Canada) Limited

Printed in Switzerland

To Willy,
who is now on the Moon

PREFACE

Sometime around 1944, when I was wrestling with recalcitrant radars in the wilds of Cornwall, a tattered issue of *Life* magazine fell into my hands. It contained pages of stunning color pictures—photographically detailed paintings—of the ringed planet Saturn, as viewed from its various moons. Although I had seen similar pictures before, these were in a class by themselves, and for a long time I had them pinned up on the wall of my billet. Not for nothing was I known as "Spaceship" through most of my Air Force career.

I had never heard of the artist, Chesley Bonestell, but thereafter I looked out for his work and collected it whenever the opportunity arose. A few years later, I was delighted to hear from my friend Willy Ley that he was writing the text to go with a book of Bonestell's paintings.

Conquest of Space appeared in 1949, and probably did more than any book of its time to convey to a whole generation the wonder, romance and sheer *beauty* of space travel. Turning its pages today, almost a quarter of a century later, I can still recapture some of that initial excitement. About half of the book has already happened; the scenes which Bonestell painted, using the best contemporary knowledge, have now been photographed by robot spaceprobes—and even more astonishing, by hand-held cameras. On the whole, the pictures stand up very well; sometimes, indeed, they are uncannily prescient, and future generations will hardly believe that they were so far in advance of the reality.

In 1970, quite out of the blue, I received a letter from Bonestell asking if I would be interested in writing the text for his latest space project. Although I had recently sworn, on a stack of vintage *Wonder Stories,* that I would hereafter write only fiction, I found it impossible to resist the invitation. It was an exciting challenge—and I felt that it was the least I could do for him, in return for the pleasure and inspiration his pictures have given me over so many years.

The result is this book. It could have been written by Willy Ley, Bonestell's longtime collaborator—but alas, Willy died only three weeks before Apollo 11 demonstrated the truth of all that he had written for more than forty years. Now he has had a crater named after him on the Moon. You will find him in the new maps of Farside, among the great astronomers and scientists of the past.

A.C.C.

ACKNOWLEDGMENTS

It is a pleasure to thank the following for their help during the preparation of this book:

William Pickering, director of the Jet Propulsion Laboratory.

Bruce Murray, professor of planetary science, California Institute of Technology.

Frank E. Bristow, James A. Dunne, Ray L. Newburn and Ronald Toms, of the Jet Propulsion Laboratory.

Robert Leighton, astrophysicist, California Institute of Technology.

Fred A. Franklin, astronomer, Astrophysical Observatory, Smithsonian Institution.

Phillip Chapman, scientist-astronaut, Manned Spacecraft Center, Houston.

And especially, Bob and Ginny Heinlein, for arranging a Grand Tour to the Bonestell studio.

C.B. and A.C.C.

CONTENTS

ILLUSTRATIONS

COLOR PLATES

(between pages 60 and 61)

Spacecraft after encounter with Saturn and now 15,000 miles from Saturn's satellite Rhea
The rings of Saturn
Uranus from Umbriel
Neptune from Triton

BEYOND JUPITER

ONE
OPENING FRONTIERS

> To us a thousand years later, the whole story of mankind before the twentieth century seems like the prelude to some great drama, played on the narrow strip of stage before the curtain has risen and revealed the scenery. For countless generations of men, that tiny, crowded stage—the planet Earth—was the whole of creation, and they the only actors. Yet towards the close of that fabulous century, the curtain began slowly, inexorably, to rise, and Man realized at last that the Earth was only one of many worlds, the Sun only one among many stars. The coming of the rocket brought to an end a million years of isolation.
>
> —*The Exploration of Space* (1951)

Twenty-one years ago, I put these words into the mouth of a latter-day Toynbee—an imaginary historian of the third millennium. At that time, I never dreamed that the first artificial satellite (Sputnik I, 1957) lay only six years ahead, and that within a decade (Gagarin, 1961) man himself would have entered space.

Still less did I imagine that, during this twenty-one-year period, the first phase of manned lunar exploration would not only have begun—but would also have ended. Project Apollo (1961–1972) refuted all the prophets who assumed that we would reach the Moon when the time was ripe, through the orderly development of space technology. Instead, a series of political events, which no one could possibly have predicted, brought astronautics into the realm of international affairs.* A tiny handful of men recommended the Moon as a goal, in an attempt to repair their country's damaged prestige; an ambitious young President accepted a challenge whose outcome he would never live to see; an aroused Congress gladly voted the necessary funds. After that, events proceeded with what now seems a dream-like inevitability, until those first flickering TV images of a human foot descending upon the soil of another world.

It could thus be argued, with some reason, that the lunar landing was premature, occurring out of the logical time sequence. David Scott, the Apollo 15 commander, has pointed out an interesting parallel with the history of Antarctic exploration. The South Pole was reached in 1911 by the most primitive techniques—dogsleds and men on foot. Yet it was not until the International Geophysical Year of 1957, almost half a century later, that men really started to *live* there. The establishment of the United States' Amundsen-Scott South Polar base was made possible only by enormous improvements in air transportation.

It requires a considerable act of imagination to compare a Saturn V to a dogsled—which, after all, *can* be used more than once. And that, of course, is the trouble with our current "throw-away" rocket technology. There will be a short pause in the exploration of the uni-

*See John M. Logsdon's *The Decision to Go to the Moon* (Cambridge, Mass.: MIT Press, 1970).

verse, while we develop reusable systems. The so-called Space Shuttle, now on the drawing board, could be the DC-3 of astronautics. Or, at least, the DC-1. . . .

Barring social and economic disasters, the large-scale, *manned* exploration of space should have reached this new level of competence at the beginning of the eighties; a half-century hiatus, like that at the South Pole, though possible, is most unlikely. Before 1990, therefore, we should have established permanent bases on the Moon, orbited Venus, and landed on Mars, as well as made a rendezvous with its tiny satellites, Phobos and Deimos. (One can hardly talk of "landing" on worlds where gravity is so feeble that the high-jump record would be a couple of miles.) We may also have visited the nightside of Mercury, as well as a few convenient asteroids and comets. All this will be enormously exciting, and will create entire new sciences, yet when plotted on a chart of the whole Solar System, it is very parochial, short-range stuff. We will still be like early Greek sailors, milling around inside the Mediterranean, while beyond the Pillars of Hercules rolls the vast Atlantic, sundering us from who-knows-what mysterious continents.

For beyond Mars, the scale of the Solar System suddenly expands. We must no longer think in tens of millions, but in *hundreds* of millions of miles—ultimately, in billions. It is true that such distances are still trivial when compared with those to the stars; nevertheless, it will take several generations of space technology to shrink them to manageable size.

Mere distance, however, is not the only challenge we must face as we head out into the cold wilderness on the other side of Mars, and the Sun shrinks to a blinding, heatless point of light. The planets that lie out there are of a totally new kind, so strange that it is probably as hard for us to envisage them as it would be for an intelligent fish to imagine conditions on land.

The four outer planets Jupiter, Saturn, Uranus and Neptune are huge—up to eleven times the diameter of Earth—but they are made of such lightweight materials that they have been given the picturesque label "Gas Giants." The name is not wholly accurate since they must have solid cores; the cores, however, may lie beneath so many thousands of miles of steadily thickening atmosphere that neither men nor instruments may ever reach them.

Since the Moon and Mars have already provided us with such torrents of information that the data banks are hard put to store it, some might argue that these remote and exotic worlds should be left to their own devices for a century or so. Perhaps this will be true, at least as far as *manned* exploration is concerned. But [illegible] spaceprobes, the situation is entirely different. By a rather remarkable (a few might even say, [illegible] coincidence, the opportunity exists in the late 1970s to reach *all* the outer planets with very little more difficulty than Mars or Venus. It is an opportunity which, if not grasped then, will not recur for 171 years. One wonders what future generations will think of us if we miss a chance which will not come again until the year 2147.

Shakespeare put it well when he wrote: "There is a tide in the affairs of men, which taken at the flood, leads on to fortune." The Solar System also has its "tides"; as the planets sweep along their orbits, their gravitational fields are constantly intermeshing—sometimes canceling, sometimes reinforcing each other. And dominating them, of course, is the far greater field of the Sun itself, which controls all movement in an area billions of miles across.

Occasionally, during this invisible tug-of-war, situations arise when certain journeys become much easier than at other times. It is not a simple alignment of planets but, as we shall see later, something much more complex. The most favorable opportunities open up—quite literally—an easy way to the stars.

That such a thing was possible could have been discovered by any competent mathematician at any time in the last three hundred years. During the nineteenth century, a few astronomers came very close to it when they studied the orbits of certain comets. They found that comets occasionally disappeared from the Solar System, as a result of an encounter with Jupiter. The gravitational field of the giant planet could, under the right conditions, produce a kind of slingshot effect, boosting the speed of a body passing close by. Thus a comet which came too near Jupiter could be accelerated right out of the Solar System, escaping completely from the Sun and, perhaps, one day becoming the captive of another star.

This was very interesting to the astronomers, but no one thought of its practical consequences until the 1960s. Then the space mission planners, with something of a shock, suddenly realized that the key to the outer planets lay within their grasp. What was true of Jupiter was also true, to a slightly lesser degree, of Saturn and the other giants. They could *all* give a free "gravity assist" to spacecraft approaching along suitable orbits.

The possibility therefore arose of arranging what was promptly christened a "Grand Tour" across the Solar System, provided the spaceprobe began the trip in 1976, 1977, or 1978. If the necessary standards of reliability and navigational accuracy could be attained, we could fly a spaceprobe to Jupiter, let Jupiter flick it on to Saturn, Saturn to Uranus, and so on. . . . It would be a kind of celestial billiards game, but with a fundamental difference: In billiards, when the ball in play cannons off another ball, it loses energy and slows down. In the right kind of planetary encounter, however, a spaceprobe would *gain* speed—often by a substantial amount.

It is one of the rare cases in which Nature appears to give something for nothing, and the more the concept was studied, the more interesting the possibilities became. Unbelievable though it seemed, within a mere twenty years of the launching of the first satellite, the entire Solar System had become accessible to man—right out to lonely Pluto, almost four billion miles from the Sun. (See Plate VI.)

In space research, as in ordinary life, it is impossible to do everything that is worthwhile; one has to decide on priorities, and this is often an agonizing business. There was never any chance that NASA would be able to conduct *all* the Grand Tour missions, and early in 1972 it was decided that the development of the Space Shuttle would take precedence over outer-planet exploration. Some limited missions beyond Jupiter — in addition to those described in Chapter 6 — may be carried out by the United States; but just which still remains to be seen. For NASA, the Grand Tour has, alas, become the Economy Tour.

But it is inconceivable that the Soviet Union — in view of its past enormous efforts at planetary exploration, culminating in the first landings on the Moon, Venus *and* Mars — will not take advantage of this rare opportunity. Though the USSR has always had an acute sense of history, it surprised many by its failure to place a man on the Moon fifty years after the October Revolution. It will indeed be ironic if owing to the laws of planetary movements the USSR's most ambitious space project is launched exactly two hundred years after the signing of the Declaration of Independence. . . .

TWO
COSMIC BILLIARDS

Stealing energy from one planet to travel on to another is a neat trick, and fairly advanced mathematics is required to explain exactly how and when it can be done. Fortunately, it is possible to get a very good idea of the principles involved without the use of any mathematics at all.

A space vehicle approaching an isolated planet from a great distance will always have some initial speed, and will steadily gain velocity as it falls deeper and deeper into the planet's gravitational field. Its greatest speed will be at the point of closest approach; then—except of course in the case of an actual impact—the vehicle will swing around the planet and head out into space again, losing speed as it departs. If the approach is very fast and very close, the orbit may be a hairpin bend, but usually the change of direction is much less drastic. In every instance, the orbit is the curve known as the hyperbola—one of the "conic sections" first investigated by the Greek mathematicians two thousand years before it was realized that they are the paths followed by all celestial bodies.* See Figure 1, page 10.

Looking at the situation from the point of view of the planet, there is no difference between the approaching and the receding halves of the orbit; they are mirror images of each other. Moreover, any speed the spacecraft picks up on the way in, it exactly loses on the way out. If it was moving at a mile a second when it was ten million miles away on the approach leg, it will be moving at precisely a mile a second again when it is once more ten million miles out, heading away. It may have whipped round the planet at fifty miles a second at its closest point, but all the extra speed will have been shed during departure. To reverse the old saying: "What comes down must go up"—and at exactly the same speed.

So in this simple case, nothing at all has been gained. All that has happened is that the spacecraft's direction of motion has been altered, perhaps by a large angle (the change of course during such a "swingby" can be a right angle or more). It is almost as if the spacecraft

*The other conic sections are the parabola and the ellipse, with the circle as a particular case of the latter.

has bounced off the planet's gravitational field. If we imagine an invisible, perfectly elastic barrier behind the planet (Figure 2) we have an exact analogy of the situation.*

This "elastic impact" model allows us to understand what happens when a real spacecraft flies past a real planet. The example we have looked at so far has been grossly oversimplified because we assumed that the planet and the spacecraft were the only objects that existed. We ignored the Sun, around which the planet is moving at a considerable velocity—Jupiter's velocity, for instance, is about 30,000 miles an hour.

So now look again at Figure 2. That "invisible barrier" should be moving, to make the model a realistic one. If it is moving upward, the spacecraft will gain velocity after the encounter; if it is moving downward, velocity will be lost. Both possibilities are of practical importance, for it is sometimes necessary to lose speed deliberately, in order to be captured by the gravitational field of a target planet.

So perhaps another good games analogy for this type of encounter would be tennis. You can smash a ball back faster than it arrives—or you can absorb some of its momentum, so that it barely crawls back over the net.

One of the most fundamental laws of nature is that energy can be neither destroyed nor created—only converted into some other form. When a spacecraft gains speed after an encounter with a planet's gravitational field, the energy that enables it to do so can only have come from the planet.

The "impact" will also have slowed the planet down, but of course by an immeasurably small amount. *How* small can easily be estimated for any practical case, like a Jupiter flyby. The weight (strictly, mass) of Jupiter is about two million million million million (2×10^{24}) tons. The effect of a spacecraft upon this would be comparable to that of a speck of dust falling on an elephant. Jupiter would drop sunward—by about the diameter of an atom. It is

*I am indebted to Maxwell W. Hunter's stimulating *Thrust into Space* (New York: Holt, Rinehart & Winston, 1966) for this analogy.

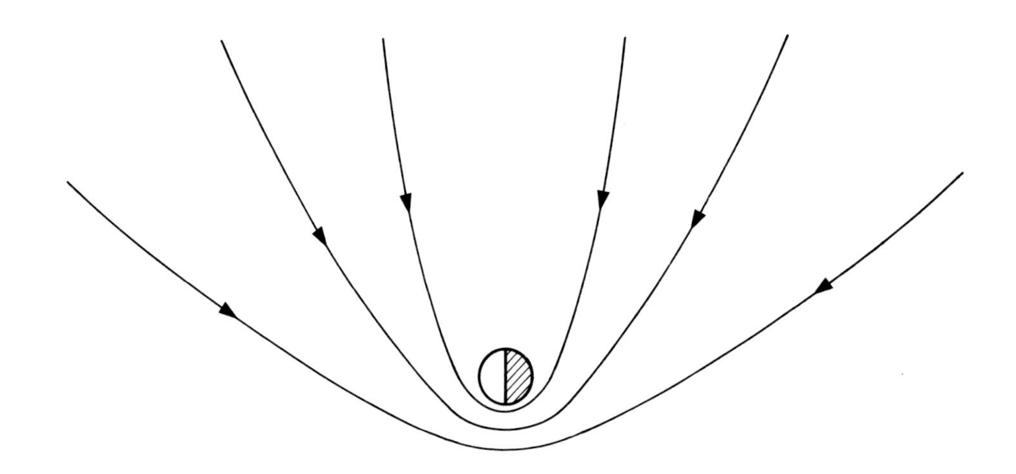

FIGURE 1.

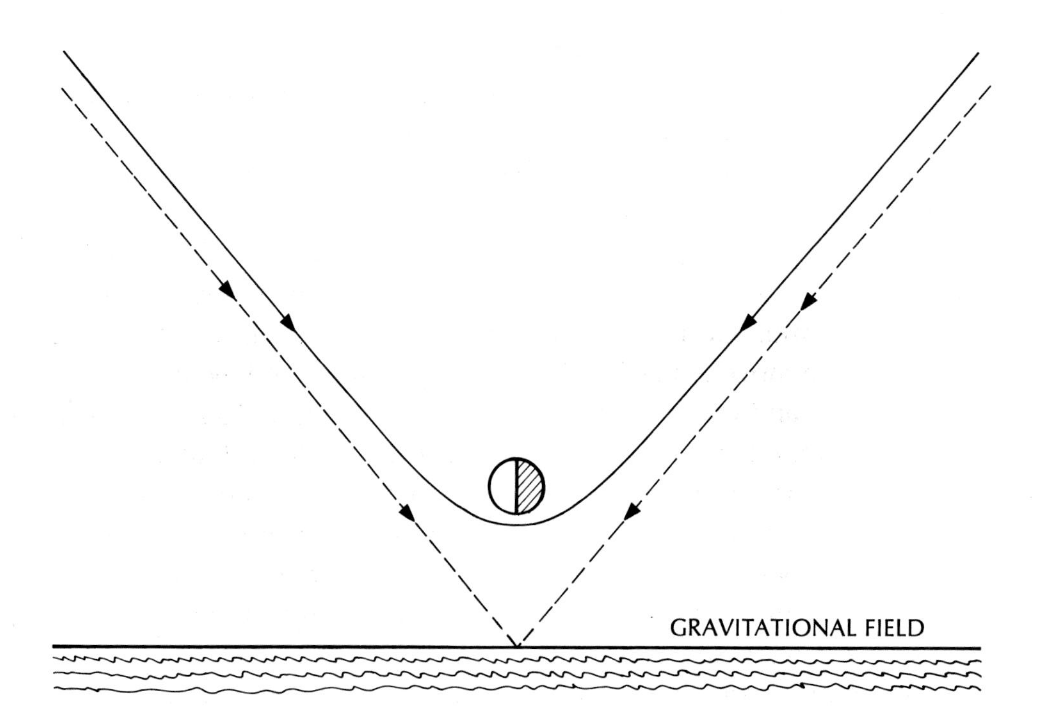

FIGURE 2.

fairly obvious that no amount of interplanetary maneuvering is going to lead, in the foreseeable future, to the collapse of the Solar System. Indeed, the natural cometary traffic that has been going on for billions of years must already have had far more effect upon the orbit of Jupiter than anything that mankind is ever likely to do.

The amount of gravity assist which one can obtain from a planet depends upon the closeness and the angle of the spacecraft's approach, and the planet's mass. For this reason, giant Jupiter is much the most effective "booster," and it is also in the right place for us to take the maximum advantage of it. It is almost as if we have a free gas station half a billion miles from the Sun, on the road to the outer planets.

But Jupiter takes nearly twelve years to complete one orbit, so there are only certain times when we can conveniently use its facilities. If we imagine a clockface with the Earth at the tip of the minute hand, and Jupiter at the tip of the hour hand, we have quite an accurate model of their relative movements. The fast-moving minute hand overtakes the hour hand approximately every hour and five minutes (12:00, 1:05½, 2:11, and so on); the fast-moving Earth overtakes Jupiter every thirteen months. Wherever the Earth and Jupiter may be at any given time, they will be back in the same relative position a year and a month later. If you miss your launch date, therefore, this is how long you may have to wait for the next, unless your rocket has the extra performance needed for the more demanding flight trajectory. In practice, there is always some margin; the "window," or interval during which a launch is still possible, is several weeks wide.

Thus Jupiter cooperates rather nicely with our calendar; we can fly to him every year, and if our navigation is accurate, he will boost our spacecraft onward and outward from the sun—toward Saturn.

This is where the situation starts to become complicated, because Saturn moves even more slowly in its orbit than Jupiter—hence its astrological reputation as "the bringer of old age." The Saturnian year is thirty times as long as ours; we will therefore have to wait many years before Earth, Jupiter *and* Saturn come into a position which allows us to carom most efficiently from one to the other.

And suppose we then want to use Saturn to boost us on to the next planet, Uranus (period, 84 years), Uranus on to Neptune (166 years) and finally Neptune on to Pluto (250 years!). It is obvious, without doing any mathematics, that the outer planets repeat their configurations only after enormous intervals of time. Strictly speaking, they *never* do so, because their orbits are not perfect circles and are tilted at various angles to each other.

Fortunately, we do not require absolutely exact lineups of the planets; even approximate configurations can be very useful, and a splendid series occurs from 1976 to 1979. During this interval it is possible to fly a number of different missions, three of which are shown in Figures 3, 4, and 5, respectively. (These diagrams may be regarded as maps of the Solar System, seen from a vantage point high above the Sun.) Figure 3 plots a Jupiter-Saturn-Uranus-Neptune mission, to which the name Grand Tour was originally applied. The term is now used more or less indiscriminately for any flight to a number of outer planets.

Although it is desirable to pass close to Saturn, to obtain the maximum boost from its gravitational field, the wide-ranging system of rings may make this dangerous. Some missions may therefore omit Saturn, and go directly from Jupiter to Uranus (Figure 4).

However, it should be fairly safe to go close to Saturn *if* the equatorial plane—in which the rings lie—is avoided. In this case, the spacecraft would fly over the north or south pole, and the resulting gravity assist would flick it right out of the general plane of the Solar System (the Ecliptic). But this is just what is desired if one wishes to go to Pluto, which has a very inclined orbit—tilted to the Ecliptic at 17 degrees. So a Jupiter-Saturn-Pluto flight is an attractive proposition (Figure 5)—especially as it would give a view of the rings unobtainable from Earth (Figure 24).

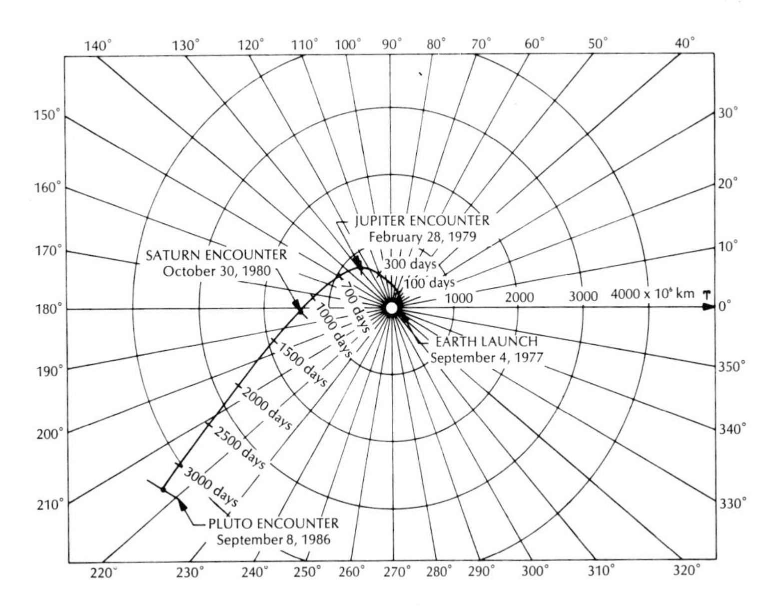

FIGURE 3.
1977 Inner Ring Grand Tour

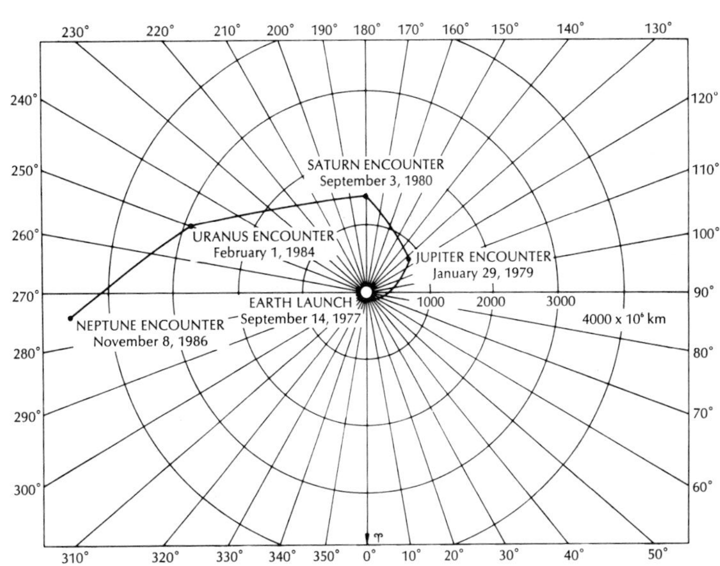

FIGURE 4.
1979 Earth-Jupiter-Uranus-Neptune Mission

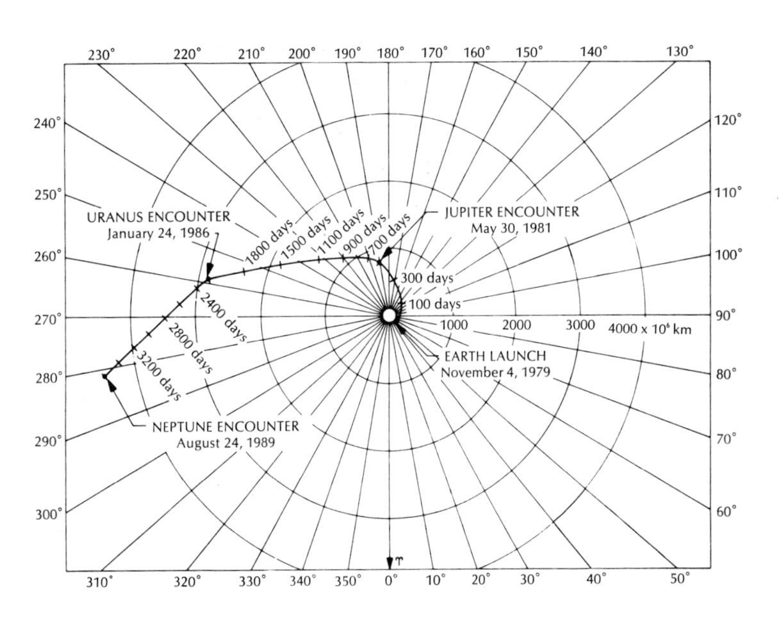

FIGURE 5.
1977 Earth-Jupiter-Saturn-Pluto Mission

Some of the most interesting of the Grand Tours are listed in the table.

Possible Missions to the Outer Planets

	Mission I	Mission II	Mission III	Mission IV
Leave Earth	1976	1977	1977	1978
Fly by Jupiter	1978	1979	1980	1981
Fly by Saturn	1980	1980	1981	–
Fly by Uranus	1984	1984	1985	1985
Fly by Neptune	1987	–	1988	1988
Fly by Pluto	↓	1986 ↓	↓	↓

(and on out to the stars)

The mathematicians of the Jet Propulsion Laboratory, Pasadena, where much of the planning for the Grand Tour has been done, have made a remarkable film of these missions. The images were all generated by a computer, which calculated the positions of the spacecraft, the planets, their moons and the brighter background stars—and then "drew" them on a cathode-ray tube, so that they could be photographed by a movie camera. The result is a striking cartoon film, untouched by human hands, which gives a vivid impression of a flight to the outer planets.

When I was at JPL in November, 1971, during the Mariner 9 encounter with Mars (see Chapter 4), I had a chance to see this film, and now feel that I have actually been on the Grand Tour. Jupiter first appears in the distance as a little sphere ruled with lines of longitude and latitude, and with its inner moons scurrying around it; even the position of the Great Red Spot is marked, by a small oval. The planet expands in the field of view, turning as it does so; on the screen, days flash by in seconds. As we sweep over the nightside, the globe suddenly becomes a thin crescent; at the moment of closest approach, everything happens very quickly. Then the planet is gone—but one of the moons swells up, appears to roll over as we pass beneath it, and also dwindles astern. . . .

The Saturn flyby is even more dramatic because the unique system of rings makes Saturn the most spectacular of all the planets. They first appear edge-on, as a thin line; then they open wide as the spacecraft flies past the planet, and close again as it recedes into the distance. Chesley Bonestell's paintings (Figures 23, 24; Plates XII, XIII) are all based on the calculations of the scientists who made this film, so one can imagine what the real thing would be like at the 1981 premiere.

Before the advent of computers that could produce such visual displays, it would have been a hopeless task to investigate all the possible types of Grand Tour. But now it is easy to "fly," in a few minutes, missions that will in actuality take ten years, and to see what adjustments must be made to make them safer or more scientifically rewarding.

Though the planets themselves are the main objective, it is obviously desirable to choose trajectories that will go close to as many as possible of their numerous satellites; anything learned about these little worlds will be a welcome bonus. It is also essential to avoid hazards, not all of which are known at this stage. Possible debris inside the rings of Saturn is one; the great radiation belt around Jupiter (the equivalent of Earth's Van Allen zone, but far more intense), which could cause damage to the spacecraft's instruments, is another. All the alternative routes can be explored quickly with the computer and displayed in a form that can be readily grasped by the eye and brain—unlike the acres of numbers which are the usual form of computer output.

It should be emphasized that all these missions can be flown with *existing* launch vehicles —and not the largest and most expensive ones. There is no need to use the huge Saturn V, or its smaller brother Saturn I; of course, if they were available, they could send really massive (multi-tonned) payloads to the outer planets. The Atlas-Centaurs and Titans (with strap-on solid boosters) are quite adequate for the task, thanks to the help given by gravity *en route*.

But the saving in energy which permits the use of modest launch vehicles is not the only benefit of gravity assist. Even more striking, and at least as valuable, is the saving in time.

As the table of the four possible missions to the outer planets shows, this technique allows us to get all the way out to Pluto in only nine years, with Jupiter and Saturn included for no extra charge. Yet a spacecraft launched *only* to Pluto by the most economical orbit would normally take forty-two years to get there! Leaving in 1977, it would arrive in 2019—instead of 1986. Even to Neptune, direct flight takes seventeen years; with gravity assist, we can do it in ten. Of course, there is a price to pay for this fantastic saving of time, but it is a reasonable one. Extremely accurate navigation will be needed, especially at the critical moments of flyby; the slightest error then could throw the spacecraft wildly off course, so that it would miss its next target planet and go wandering off into the wrong part of the Solar System. (Even so, it would still send back useful information.)

It will also be necessary to build spacecraft with a degree of reliability never before required, if they are to function for ten years or more. But we have learned a great deal since the first Sputnik was launched; there are spacecraft in orbit now that have been operating continuously for more than six years, and will go on working for many years yet.

Space is a benign environment—at least for robots. This may seem surprising, but anyone who has ever had to design equipment to work in the humid tropics, or at the bottom of the Pacific, would undoubtedly agree. It is very rare indeed for a spacecraft to be disabled by an outside force; when anything goes wrong, the fault is usually internal. And, as we shall see in Chapter 6, the probes that will be launched to the outer planets will even be able to deal with this problem. For they will be among the first of a new species of robots—those who can repair themselves.

THREE
THE FIRST SCOUTS

It would be impossible to make serious plans for missions covering billions of miles, and lasting many years, without a vast body of previous experience in space technology, much of it acquired the hard way. Behind the Grand Tour concept lie several generations of increasingly more sophisticated—and reliable—spaceprobes: Ranger, Mariner, Orbiter, Surveyor. . . .

Although the earlier writers on astronautics were quite correct in their ideas about propulsion, structures, fuels, and the main engineering aspects of space flight, they could not anticipate the electronic revolution, which made it possible to perform feats surely beyond their wildest dreams. Thus when Robert Goddard, in his classic Smithsonian paper *A Method of Reaching Extreme Altitudes* (1919) discussed the problem of proving whether or not a rocket had actually reached the Moon, the only solution he could suggest was that it should carry a few pounds of flash powder, to be ignited on contact. (He calculated that less than fourteen pounds would be strikingly visible in a 12-inch telescope, against the unilluminated surface of the Moon.) When Luna II performed this feat just forty years later (on September 13, 1959), it was followed by radio all the way, and the increase in its velocity right up to the very second of impact was accurately measured by the Doppler effect* of the returning waves.

The technology of tracking spacecraft by radio was first developed for the Vanguard satellite in the late fifties; the ability to get information back from distant, moving objects is very much older. The art of "telemetering" goes back to the weather-balloon "radiosondes," with which meteorologists started exploring the atmosphere in the 1930's. These skills, and many others, were already at hand when the time was ripe to send probes to the planets, but they would not have been practical without another development—the rise of solid-state electronics. The transistor and its still smaller successors started to replace the bulky, fragile and power-consuming vacuum tube at just the right time.

*Familiar from the change in pitch of a locomotive whistle as a train passes by. A similar shift of frequency occurs with radio waves, and gives an extremely sensitive measurement of velocity.

Electronic equipment that, in the forties, would have filled a room, had shrunk to shoebox size by the mid-sixties. By the seventies, it was no larger than a matchbox—and was still contracting. It must be admitted that the demands of long-range missilery, rather than the search for knowledge, had wrought these miracles; but they were there, waiting for the space scientists to use them when they were required.

The first objective was, of course, the Moon. (We are really very fortunate to have a large, interesting heavenly body so close at hand; how discouraging to be a Venusian* astronaut, with the nearest land twenty-five million miles away—a hundred times the distance of the Moon!)

The robot exploration of our single natural satellite began in 1959 with the USSR's Luna III, which took the first photographs of the hidden farside of the Moon. Although they were very crude, they gave astronomers a tantalizing initial glimpse of a land that had once been the very symbol of everything that could never be known. Surprisingly, it was six years before a second view of the farside was obtained, from the Russian "Automatic Space Station" Zond III (July 1965).

Before then, however, a series of nine Ranger spacecraft had been launched by the United States, in an attempt to learn something about the fine details of the lunar surface. After a heartbreaking series of six failures, Ranger 7 (July 1964) worked perfectly, and televised more than four thousand pictures back to Earth before it crashed into the Sea of Clouds. The last picture was taken from a distance of less than half a mile, and brought the Moon a thousand times closer than it could be observed through any Earth-based telescope.

Rangers 8 and 9 (February and March 1965) brought the project to a triumphant conclusion; it was during the final mission that the dramatic words LIVE FROM THE MOON appeared for the first time on television. The last photographs transmitted by Ranger 9 before it destroyed

*The classicists have had a field day over this. Hesperian, Cytherean, Venutian have all been proposed. Venusian is as bad as Earthian, but we are probably stuck with both.

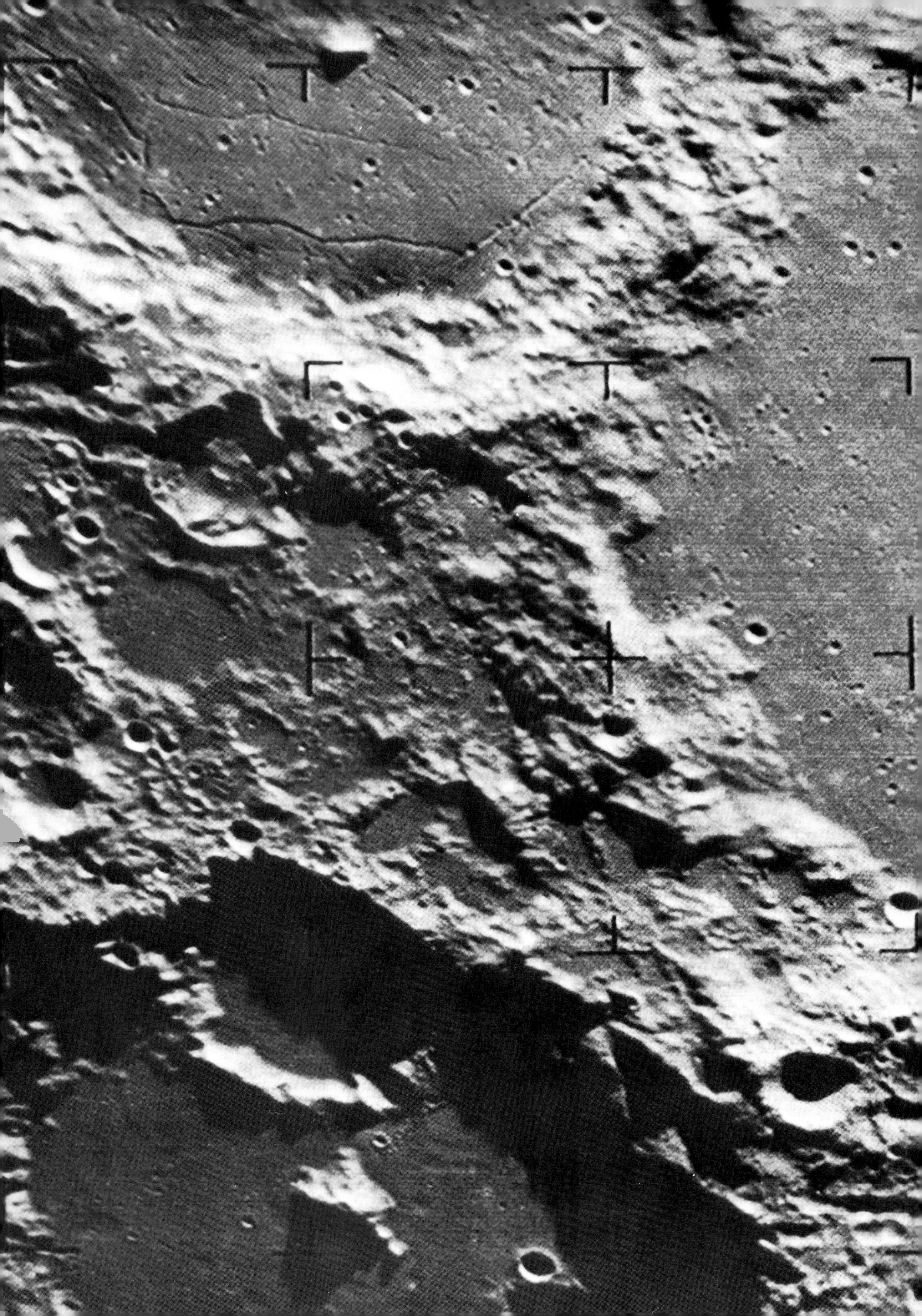

FIGURE 6. Ranger 9 television picture of the Moon taken 9 minutes and 18 seconds before impact on March 24, 1965, at an altitude of 775 miles. Slightly more than half of three major craters are shown: Alphonsus at the top, with a rille system and a central peak which rises 3,300 feet above the crater floor; Ptolemaeus at the right, without a central peak; and Albategnius at the bottom, with a central peak rising 4,500 feet.

itself on impact showed objects less than a foot across. (Figure 8). Only a year earlier, the limit of resolution for Earth-based telescopes had been a mile.

And a year later, there was another huge advance, when Luna IX and Surveyor I (January and May 1966) made the first soft landings on the Moon, and their cameras showed the detailed, closeup texture of the surface. Their touchdowns demolished many theories and exorcised many imaginary perils—particularly the long-feared lunar dust, into which it was once confidently predicted that spacecraft would sink without trace.

The Rangers, Lunas and Surveyors could examine only a limited region of the Moon, and what was also needed was a global mapping project which would cover the entire surface from pole to pole. To do this would obviously require a camera in orbit.

The USSR tackled the problem with its Zond series of spacecraft. Zond V (September 1968) was the first object to circle the Moon and return safely to earth. It carried an aerial mapping camera—*not* a TV camera—so the film was recovered and processed, as in a normal air reconnaissance mission.

This technique, repeated in later Zond flights, has a good deal to recommend it because there is nothing to match the quality of an original photograph. But since it demands physical recovery of the spacecraft (or at least of the film capsule) it has severe limitations, and obviously cannot be used for long-range interplanetary missions. In such cases, all information—scientific data, instrument readings, pictures, spacecraft performance figures—has to be sent back by radio.

Between August 1966 and August 1967, the United States launched a series of five Lunar Orbiter spacecraft—which among them photographed virtually the entire surface of the Moon—front and back—at ten or more times the resolution obtainable from the best telescopes on Earth. So climaxed three hundred years of patient effort, by thousands of amateur and professional astronomers, to map the face of the Moon. This is not to say that ground-

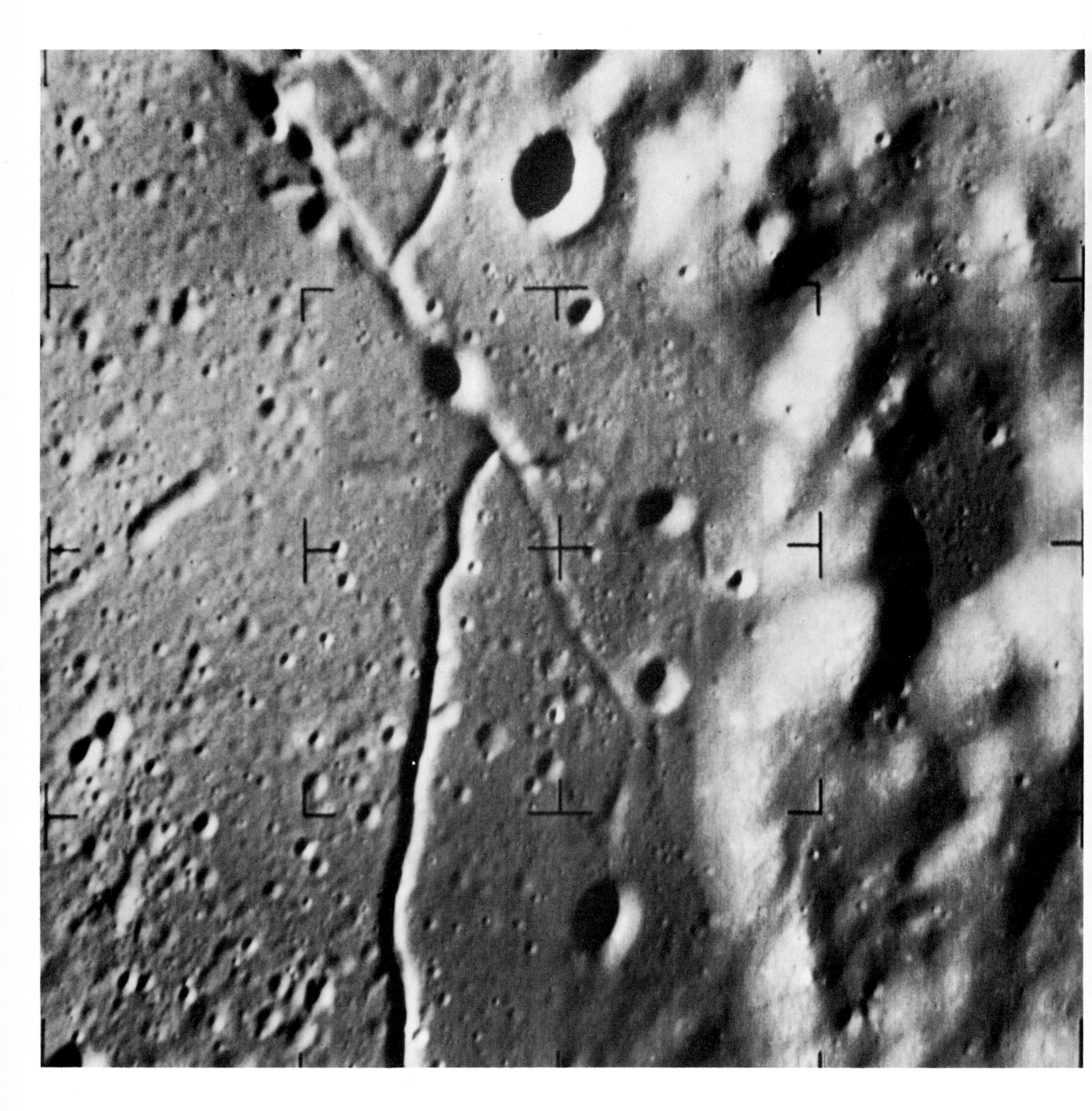

FIGURE 7. Ranger 9 photograph taken 1 minute and 12 seconds before impact in the Crater Alphonsus. Altitude above lunar surface: 107 miles.

FIGURE 8. Ranger 9 Photograph taken 2.97 seconds before impact. Altitude above lunar surface: 4.5 miles. The impact point on Alphonsus is marked with the white circle. The smallest craters visible are forty feet in diameter.

FIGURE 9. The crater Copernicus (60 miles in diameter and 2 miles deep) is shown in fine detail in this photograph of the Moon taken by Lunar Orbiter II on November 23, 1966. The mountains rising from the flat floor of the crater are about 1,000 feet high. From the horizon to the base of the photograph is about 180 miles.

based observations are no longer of any value, and the recording of surface details remains an endlessly fascinating hobby. Indeed, it could be argued that observation of selected regions by experts has become even more important—and may be much more rewarding—now that they know exactly what they are looking at. Until there is a *permanent* orbital patrol of the Moon, there is still no substitute for Earth-based observers.

The superb Orbiter photographs were sent back to Earth by a combination of photographic and electronic techniques known as "film readout." The images of the Moon were photographed on a long roll of special film, which was processed aboard the spacecraft. When the film had been developed and fixed, the images were scanned, line by line, by an extremely small spot of light. A photocell "read off" the reflected light, and the resulting signal was radioed back to Earth, where it was a straightforward matter to reconstruct the original image.

Among the highlights of the Orbiter missions were the famous photograph of Copernicus—the first portrait of the lunar world—and the mapping of the sites for all the Apollo landings. The astronauts would never have been able to make their exploring trips, especially the long-distance traverses in the Lunar Rovers, without the beautiful photographs provided by the Orbiters. (See Figure 9.)

While the Moon was disgorging its secrets, spaceprobes had also been launched to the two nearest planets, Mars and Venus. On December 14, 1962, Mariner 2 became the first spacecraft to radio back information from another planet when it flew past Venus at a distance of 22,000 miles. It carried no picture-taking equipment, but it obtained valuable scientific information and confirmed the hypothesis—which many scientists had been reluctant to subscribe to—that Venus is extremely inhospitable (see Chapter 5). The last doubters were convinced when the USSR's Venera 4 succeeded in dropping an instrument-carrying capsule into the planet's atmosphere on October 18, 1967; it was crushed by the unexpectedly high pressure while still some twenty miles above the surface.

Venera 4 had been preceded by at least eleven failures; the Russian deep-space effort has been far more ambitious than the American one, and started much earlier, with one probe being aimed at Mars as early as October 1960. It was not until December 1970 that Venera 7 penetrated the full thickness of the atmosphere and broadcast from the almost red-hot surface of Venus for twenty minutes, before succumbing to heatstroke.

However, Mars had always been regarded as a more promising objective, especially from the photographic point of view. Unlike Venus, it was not perpetually shrouded in clouds; its thin atmosphere (incorrectly believed, before the 1960s, to have about one-tenth the density of Earth's) seldom obscured the surface details which had intrigued and tantalized astronomers for more than a hundred years.

The first TV camera was flown past Mars on July 14, 1965, aboard Mariner 4, and radioed back twenty-one pictures of quite limited resolution. Nevertheless, they were full of surprises, and overnight most existing theories about Mars became obsolete. There was no sign, alas, of the famous "canals," and at first sight the planet which had been the home of so many romantic fantasies seemed as cratered and lifeless as the Moon. Mariner 4 also established that the Martian atmosphere was ten times thinner than anyone had imagined—only about one hundredth the density of Earth's. In its more modest way, Mars looked almost as unpromising as Venus, though not quite so ferociously hostile.

In July and August 1969 two considerably more advanced spacecraft, Mariners 6 and 7, obtained many more photographs and mapped a sizable fraction of the planet during the few hours of their flybys. But it was clear that—as in the case of the Moon—what was wanted was a spacecraft that could remain in orbit, making observations for weeks or months. This was the only economical way to study the planet as a whole and to monitor any variations that might be taking place. For Mars, unlike the Moon, is a dynamic world, with weather, seasonal changes, and waxing and waning polar caps.

The new era of Martian observations began in November 1971, with the first interplanetary orbiter—Mariner 9. (There should have been a Mariner 8, but it had been destroyed a few minutes after launch by the failure of a component, too small to be visible to the naked eye, in one of the control circuits.) Mariner 9 made a perfect lift-off on an Atlas-Centaur from Cape Kennedy on May 30. For more than five months it coasted, like an independent planet of the Sun, along the arc of a great ellipse; and on November 13, it overtook Mars.

FOUR
APPOINTMENT WITH MARS

The Ranger, Orbiter and Mariner projects have all been supervised by the Jet Propulsion Laboratory of the California Institute of Technology, and some of the greatest scientific triumphs (and disasters) of the space age have been witnessed in its Mission Control Center. On the afternoon of November 13, 1971, a number of scientists and journalists gathered in the visitor's viewing room to witness the Mariner 9 encounter with Mars.

The weather that day was perfect. For once, the Pasadena hills were visible; there was not a trace of the infamous Los Angeles smog, and we could see with crystal clarity for thirty miles. In a few more hours this was to be the subject of ironic comment.

Like the gallery of a theater, the viewing room overlooked the Mission Control Center and was separated from it by a large, soundproofed window. The general layout of the Center, with its flickering TV monitors, glowing lights, digital clocks counting seconds up *and* down, and large visual displays, was very similar to the Mission Control Center in Houston, but the scale of things was slightly smaller. The knowledge that men's lives were not involved reduced the tension, but the feeling of suspense was still considerable. At stake were more than a hundred million dollars, countless thousands of the most skilled man-hours in the world—and the only chance, for several years, to multiply at least tenfold all that had ever been learned about Mars in the centuries since men had studied the planet. Probably no one had forgotten the fate of Mariner 9's precursor, now attracting inquisitive fish somewhere on the Atlantic seabed.

But Mariner 9 had now covered 248 million miles in five months, and was still operating flawlessly. Forty-eight hours earlier, while the spacecraft was some half-million miles from the planet, its two TV cameras had been switched on, and their pictures of Mars had been played back to Earth. They had shown, as expected, a small, featureless disk, like the Moon about three days from full, so it was known that the camera system was operating normally.

The large TV monitor screens in the viewing room now displayed rows and columns of numbers which, to the initiated, gave a complete analysis of the spacecraft's state of health. Dozens of readings of temperature, voltage, gyro angles, propellant pressure and other more esoteric quantities were being continually taken, converted into digital pulses, and beamed back to Earth with a power of only twenty watts—about that of the feeblest lightbulb. Seven minutes later, traveling at 186,000 miles a second, an almost infinitesimal fraction of that energy arrived at the great 210-foot tracking antenna at Goldstone, California.

Amplified, cleaned up, and decoded, the pulses were converted into numbers and displayed on the monitors, thus telling the engineers everything they needed to know about Mariner 9, now so far from its builders. Most of these numbers did not change for hours at a time, because the spacecraft was still in cruise mode, its energies stored up and waiting for their moment of release.

Mariner 9 was racing in toward Mars, still accelerating in the planet's gravitational field, at over 11,000 miles an hour. If unbraked, the probe would make a right-angle turn around Mars at a height of some 800 miles, and then go into orbit around the Sun. To become a third moon of Mars, it would have to reduce its speed by 3,600 miles an hour; otherwise the planet could never capture it.

About two hours before the encounter with Mars, Goldstone had beamed the sequence of pulses which told the spacecraft's computer, "Initiate maneuver sequence." The signal was repeated four times—just to make sure that it was heard. If, for some reason, there had been a breakdown in communications with Earth, the central computer would have gone ahead anyway at the correct time. All essential instructions had been given to it in advance, but Mission Control could change them if necessary. Mariner 9 had to have a great deal of independence since it would take seven minutes for any orders to reach it—and the ground controllers could not know for another seven minutes if they had been obeyed.

During the next two hours, the spacecraft made its preparations for the rendezvous. The

FIGURE 10. The Jet Propulsion Laboratory, Pasadena, California.
FIGURE 11. The control room at the Jet Propulsion Laboratory.

autopilot was switched on; the tiny gyros—which measured the vehicle's orientation in space, and told it where it was pointing at any moment—were given ample time to build up speed. All the electrical systems warmed up, drawing power from the four solar panels, which made the spacecraft look rather like an old-fashioned windmill. They could extract about 500 watts of energy from the sunlight falling on them; when the vehicle went into shadow behind Mars, it had enough reserve battery power to keep operating during the period of eclipse.

At 3:52 P.M., Pacific standard time, the lines of numbers on the Mission Control monitors suddenly started to change. Seven minutes earlier, out at the orbit of Mars, the roll jets had fired.

An observer flying along with the spacecraft would not have noticed the feeble puffs of nitrogen gas from the thimble-sized micro-rockets at the tips of the solar sails. But he would have seen that the windmill was slowly turning—very slowly—taking four minutes to rotate through less than half a right angle, as if its vanes were responding to a barely perceptible breeze. However, they soon came to rest again.

Eight minutes later, the entire windmill started to slew around, as if searching for a more favorable compass bearing. This time, it swung through more than a right angle, and took twelve minutes to reach its new position. But the vanes, with their thousands of solar cells glinting like rubies in the sunlight, still refused to turn.

What the spacecraft had done was to orientate itself so that its main rocket motor was pointing in the direction of flight, while its radio antenna was aimed back at Earth. It was now in the correct position for retrofire; eight hundred miles below, the surface of Mars was rolling backward at 11,000 miles an hour.

For the scientists and engineers at JPL, now was the moment of truth. The propulsion system was a new one, never before tested in space; its rocket engine must run flawlessly for sixteen minutes, consuming almost half a ton of the reactive liquids nitrogen tetroxide and

FIGURE 12. The radio antenna, 210 feet in diameter, at Goldstone, California.

methyl hydrazine. Though these are not the most powerful of fuels, they possess one overwhelming advantage for a mission such as this: they burn spontaneously on contact, so there is no need for vulnerable ignition systems or pyrotechnics to set them off.

It was now very quiet in the viewing room; all of us were watching the monitors, and conversation had stopped completely. From time to time I glanced at Dr. Pickering, who appeared remarkably relaxed and cheerful. I wondered if he was thinking of Ranger 6, which had flown a perfect trajectory all the way to the Moon—with a camera system that turned out to be dead when it arrived. . . .

Suddenly, there was a ripple of cheers and an outburst of clapping. At 4:24 P.M. (PST), the line of telemetry numbers indicating motor chamber pressure had started to flicker. The motor was firing, delivering its full three hundred pounds of thrust. We could relax then—partially. The propulsion system still had to operate for sixteen minutes, cutting Mariner's speed by almost a mile a second, before the spacecraft could enter the desired orbit around Mars.

The telemetered figures on the monitors continued to bring in good news; the thrust was steady, the gyros were holding the correct attitude. Now, no one could believe in the possibility of failure; even though the spacecraft was not yet in the correct orbit, it had lost so much speed that Mars was bound to capture it. *Some* kind of reconnaissance could be carried out, and in the back rooms of Mission Control, all sorts of emergency and contingency plans were happily discarded.

At 4:20 P.M., precisely on time, chamber pressure and thrust dropped to zero as the motor ceased firing. Mariner 9 was in a long, elliptical orbit which would take it around Mars in just over twelve hours; later, the orbit would be trimmed by a short burn which would make the period exactly 11.98 hours. This odd figure had been chosen to suit spacecraft photography; every seventeen Mars-days, the cameras would look down on the same areas, at the same sun elevation. Only in that way would it be possible to ensure that any changes in appearance were real, and not due merely to varying shadows.

FIGURE 13. Mars from its inner moon, Phobos, 3,700 miles above the equator.

Mariner would start to take its first photos from orbit early next morning (Sunday), while it was making a pass over the south pole. At the rate of one every forty-two seconds, it would store thirty-three pictures in its little video-tape recorder, and would beam them back to Earth about an hour later. This first playback, however, was going to be somewhat tedious: each picture would require three-quarters of an hour to come through—six times longer than its coded pulses would take on the trip from Mars to Earth! During its transmission, therefore, every picture would be spread out along a band 45 light-minutes, or 500 million miles, long. Even this slow rate of acquisition was a great improvement on the performance of the first Mariner to reach Mars; then, it had taken an agonizing eight hours to build up every picture.

Now, thanks to greatly improved electronic techniques, the eight hours had been cut to either forty-five minutes or six minutes, depending on the Earth station in use. Only the huge 210-foot dish at Goldstone could collect enough power to permit the six-minute rate of reception; when the spin of the Earth no longer allowed the Goldstone dish to be pointed at Mars, smaller 65-foot dishes in Spain, South Africa and Australia had to be used, and these required forty-five minutes per picture. Mariner would always be visible from at least one of these radio telescopes—the heart of NASA's Deep Space Network.

Goldstone was back in action by 2:00 P.M. on Sunday afternoon, and the pictures began coming in at the high rate. As each frame started to arrive, every line of the scan would be counted, and the corresponding line numbers 1, 2, 3 . . . began to appear on the monitor. There were 700 lines to a scan, so excitement would mount toward the high 600's, when each new image was almost due. At 700, the monitor would flash "Line Scan Complete" and go blank for a few seconds, while the computer did the final assembly of the image. And then Mars would flash on the screen.

It must be admitted that those first images were a considerable disappointment. By a very odd—some thought suspicious—coincidence, Mars had become as coy as Venus, just when we were about to have our closest look. The entire surface of the planet was covered with cloud; one of the greatest storms ever observed was raging over the daylight side. Someone in Mission Control put up a cartoon showing the Martians sucking the still strangely absent Los Angeles smog across space, and using it to camouflage their world.*

This was very frustrating to the geologists and cartographers, and still more so to the news networks, who wanted to have some good pictures. But in the long run—assuming that Mariner functioned for its planned ninety-plus days—it would probably be a piece of luck. As they watched the storm die away, the meteorologists would learn much more about the behavior of the Martian atmosphere than if all had been calm and peaceful. And how fortunate that the earlier, flyby Mariners 4, 6 and 7 had not arrived at such a time! They had had only a few hours in which to make their observations; *this* Mariner could afford to wait for weeks.

On some images, distinct dark patches appeared—perhaps mountain peaks jutting above the cloud layer. There was a considerable amount of detail that just eluded the eye, and this could be exaggerated by "computer enhancement." Just as, on a TV set, one can produce a falsely dramatic image by turning the Contrast all the way up, so the computers could process the image—dot by dot—revealing contours of light and shade which the naked eye could not possibly perceive. The result had to be studied with caution; sometimes the enhancement put in things that simply weren't there.

Since each picture embodied five and a half million "bits" of information (700 lines, 832 picture elements per line, 9 bits per picture element), the computation involved in this was awesome. In fact, the entire mission would have been utterly impossible before the advent of high-speed computers, for during its orbital operations Mariner demanded some

*A document was later circulated at JPL in which the Martian equivalent of the CIA claimed credit for this obscuration. Authorship has been attributed to me. Like any good spook, I deny everything.

FIGURE 14. (OVERLEAF) Twelve miles above Mars, 15 degrees south on the zero meridian. This crater was first photographed in detail by Mariner 6 and is 24 miles across and 1½ miles deep.

36 *billion* calculations a day. With old-style desk machines, this would have required one quarter of the entire labor force of the United States.

Even if the television experiments had to bide their time, the other measurements were already coming in splendidly. Mariner carried three other instruments—one spectrometer for the ultraviolet, another for the infrared, and an infrared radiometer (or detector) which would map the temperature of the planet with great precision.

The spectrometers should produce a wealth of information about the composition of the Martian atmosphere and surface; the radiometer should locate any hot spots, which could be a sign of internal activity, and might suggest promising places to look for life. All three instruments were pouring back torrents of raw data and covering yards of recording paper with hen tracks which conveyed meaning only to specialists; and even the specialists would need months to interpret them.

I managed to gate-crash one of the first meetings in which all the principal scientific investigators had gathered to exchange and interpret their results, and to make plans for the next day's operations. It was an impressive collection of talent, covering many disciplines and including one Nobel Prize winner (Dr. Joshua Lederberg). The atmosphere was one of good humor and excited anticipation.

Although I understood almost a tenth of what was going on, I cannot report on it; that is the rightly, and jealously, guarded privilege of the researchers concerned. There is an elaborate protocol for the publication of scientific results; anyone who jumps the gun and makes a premature disclosure is liable to be expelled from the club. It has happened.

Three days after Mariner had gone into orbit, the storm was still raging and only a few vague glimpses of the Martian surface had been obtained. But two weeks later, the TV cameras had their first chance of showing what they could really do.

From time to time, as Mariner raced around its twelve-hour orbit, it came within a few

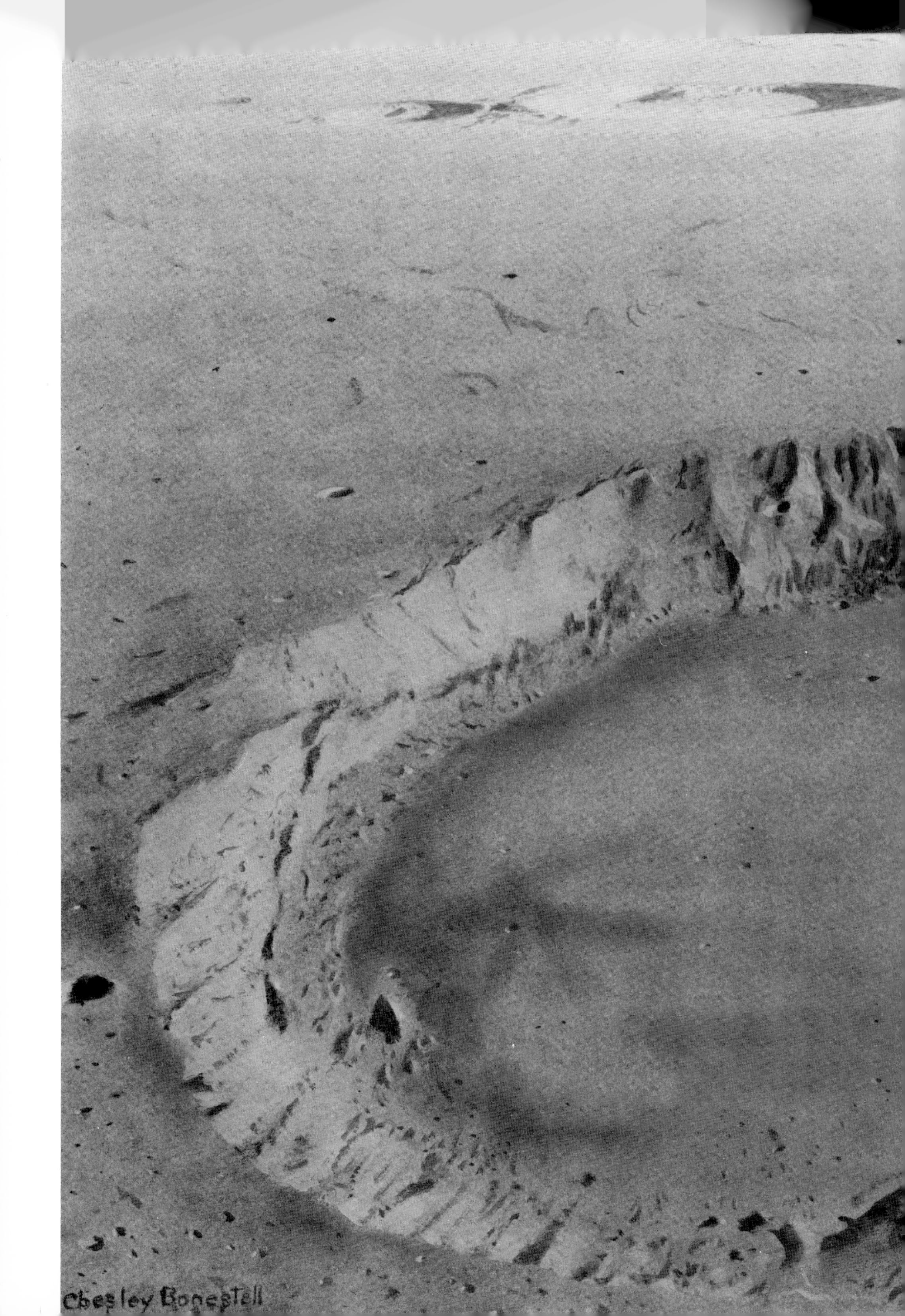
Chesley Bonestell

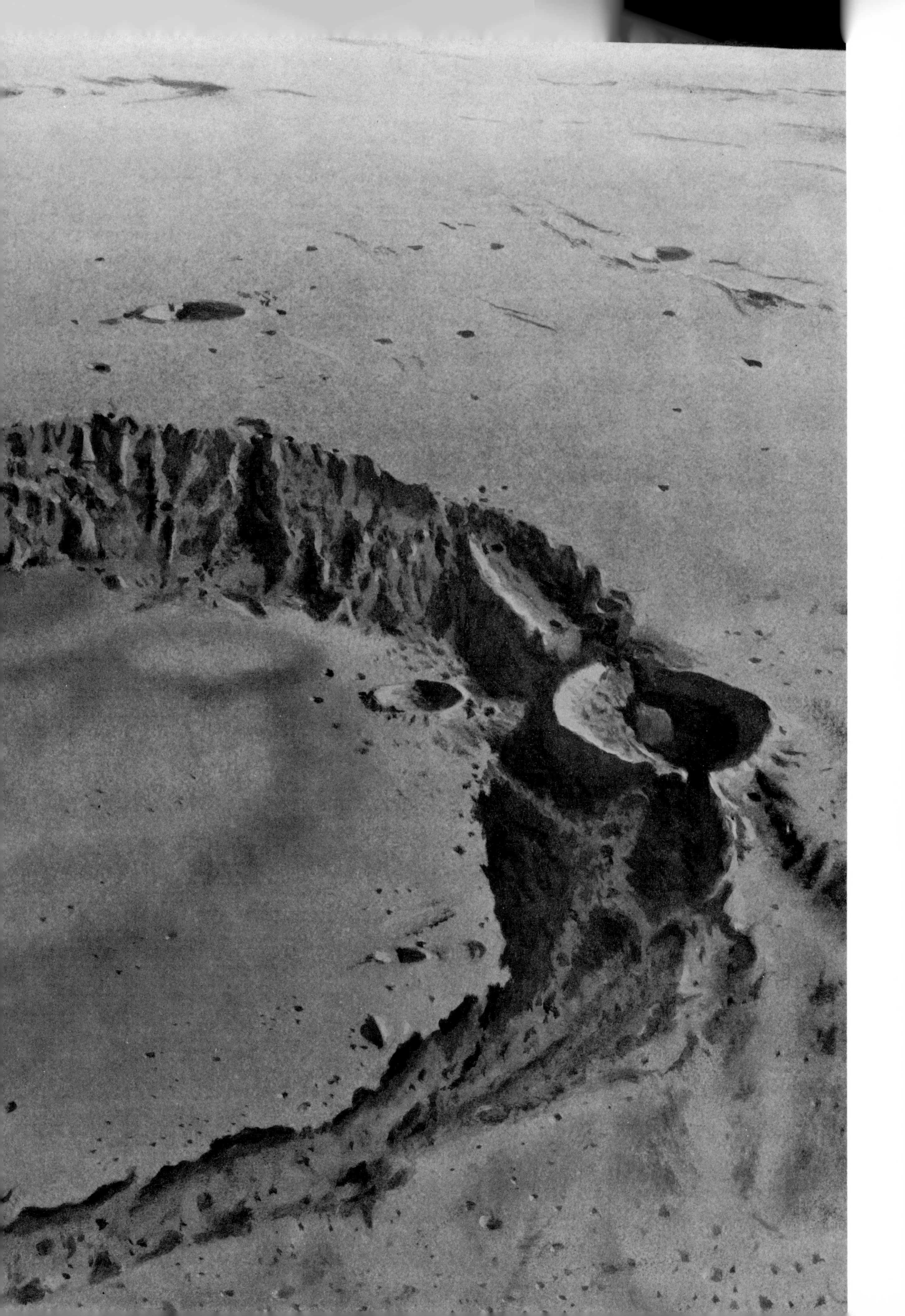

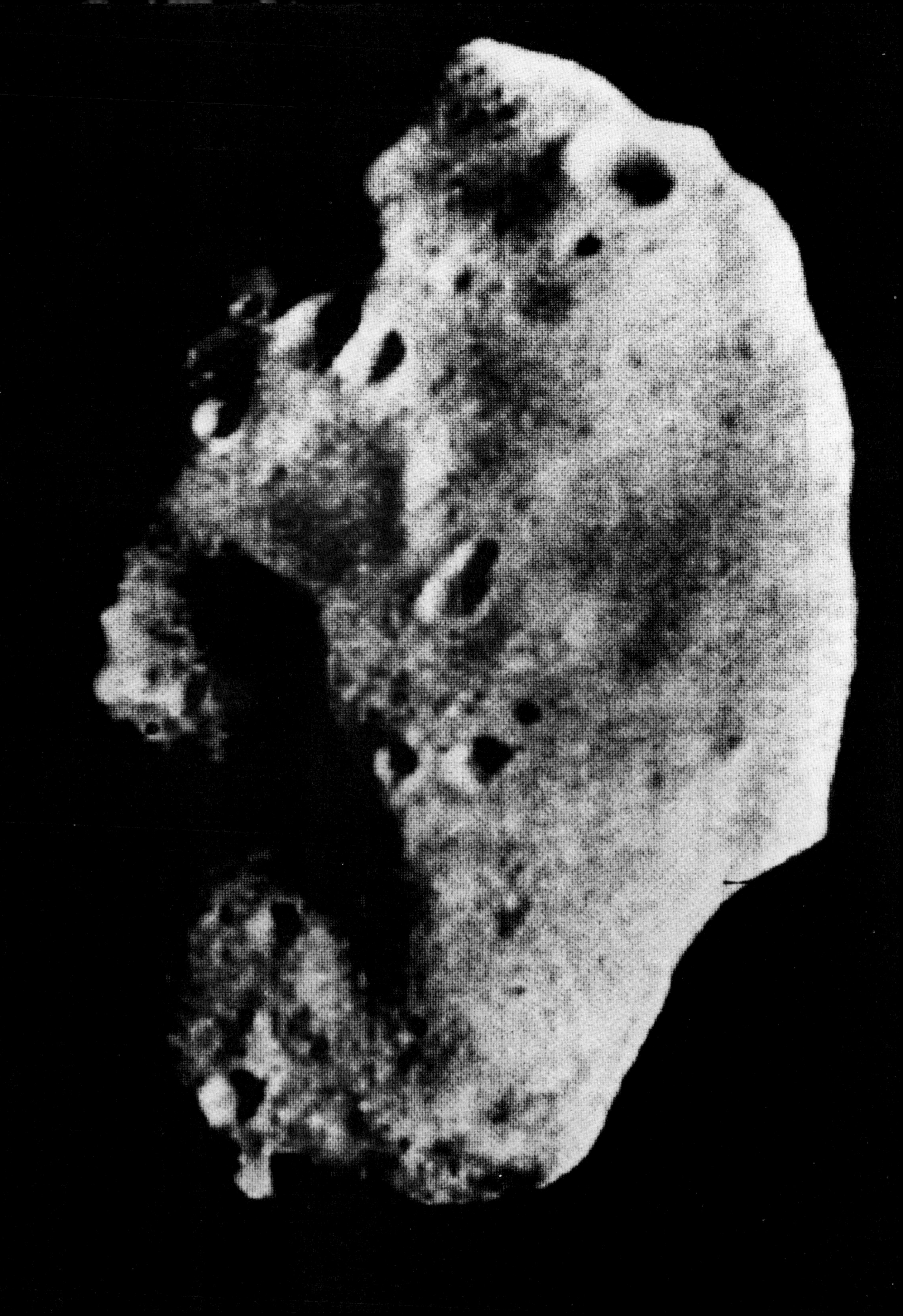

FIGURE 15. The most detailed image of Phobos available to date is seen in this computer-enhanced photograph taken by Mariner 9 during its thirty-fourth orbit of Mars.

thousand miles of the tiny moons, Phobos and Deimos. In Earth's largest telescopes, they appear no more than points of light, and were assumed to be irregular chunks of rock ten or fifteen miles in diameter. Mariner proved that this assumption was indeed true, showing the larger moon, Phobos, in amazing detail from a distance of 3,444 miles (Figure 15). Though not a very beautiful object, Phobos provided the first glimpse that man had ever had of a new type of heavenly body . . . perhaps part of the debris left over from the formation of the Solar System.

And one day, Phobos may be very important. There are a few billion tons of building material there, conveniently orbiting at the approaches to Mars. This little offshore island may be a meeting place for the commerce of the planets, in the centuries to come.

FIVE
SUNWARD HO!

According to present plans, the first spacecraft to use gravity assist will not head out toward the giant planets, but will go sunward. It will be the last of the Mariners, and its goal will be the two inner planets, Venus and Mercury. Launched from Cape Kennedy in October 1973 by an Atlas-Centaur, the half-ton spacecraft will pass within 3,000 miles of Venus in February 1974 (Figures 16 and 18). As it flies by, it will take several thousand photos of the planet at the rate of more than one a minute; the camera system will be very similar to that of Mariner 9. Details less than a mile across will be resolved—if any are to be seen in the brilliant white clouds which completely cover the planet.

The spacecraft will also carry instruments which will map the temperatures over the whole planet, and others which will analyze the atmosphere. They may help to settle one of the most baffling problems in Solar System studies: Why are Earth and Venus, which are so similar in size and distance from the Sun, completely different in every other respect?

It is only during the last few years that we have realized the extent of that difference. Because Venus is some 25 million miles sunward of Earth, it was reasonable to assume that it would be considerably warmer, but the optimists argued that the virtually complete cloud cover would reflect some of the excess heat, so conditions would be no worse than tropical. At the poles, it might even be temperate by our standards.

The dazzling white clouds "obviously" indicated vast quantities of water, so perhaps Venus was covered by huge oceans. All this sounded very much like the ancient Earth, in the days of the giant reptiles. Thus it became almost an article of faith that Venus was a primitive but prolific world, teeming with the sort of life which would make ideal trophies for intrepid big-game hunters.

This mythical Venus started to disappear as soon as the astronomers could penetrate the eternal overcast. The clouds are completely opaque to visible light, so it had never been possible even to discover the length of the planet's day—there were no landmarks to reveal

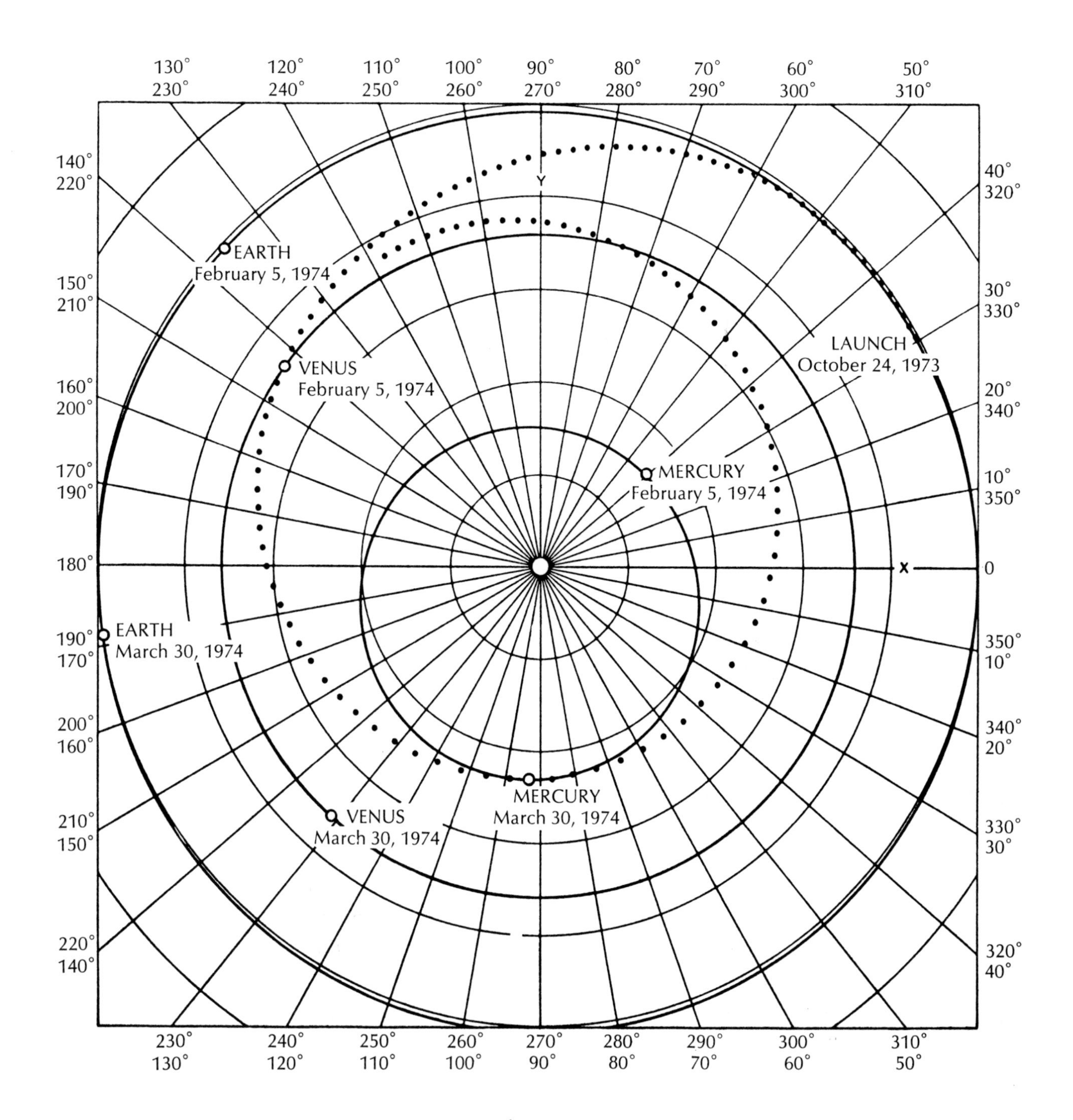

FIGURE 16. 1973 Earth-Venus-Mercury Mission

its rotation. But radio waves could go through the atmosphere unhindered, and they brought back surprising news.

The first news was that Venus is really hot—around 800°F. The second was that it rotates very slowly and, like Uranus, in a "backward" direction. The "day" of Venus lasts 243 Earth-days. And since Venus takes 225 Earth-days to orbit the Sun, we have the peculiar situation of a planet whose day is longer than its year!

The time the Sun takes to go around the Venusian sky—the solar day—is something else again; it works out at about 118 Earth-days. But this may not be of much importance to anyone on Venus, for it seems unlikely that much sunlight can penetrate to the depths of the cloudy, extremely dense atmosphere. The atmospheric pressure on the planet's surface is probably a hundred times higher than on Earth—equivalent to the pressure half a mile down in the ocean—and is entirely due to enormous quantities of carbon dioxide, present in quantities 70,000 times greater than in our atmosphere. Any other gases—oxygen, water vapor, nitrogen—make a negligible contribution. If it became cold enough on Venus for the water to condense, it would produce a layer less than a foot thick. So much for the teeming oceans. . . .

And so much, it would seem, for any possibility of life. But there is some faint—*very* faint—hope.

We have now begun to explore Venus by radar, and although the first maps are extremely crude, they appear to indicate the presence of mountains (Figure 17). Because Venus has no axial tilt, it presumably has no changes of season. No one really has any idea of the temperature variations on such a weird planet; so optimists may still hope that if there *are* high mountains at the poles, the thermometer may drop below the boiling point of water, allowing lakes to form and some kind of life to arise.

Another, even remoter, possibility is the existence of aerial life-forms, perhaps microscopic in size, in the higher and cooler levels of the atmosphere. This idea was developed

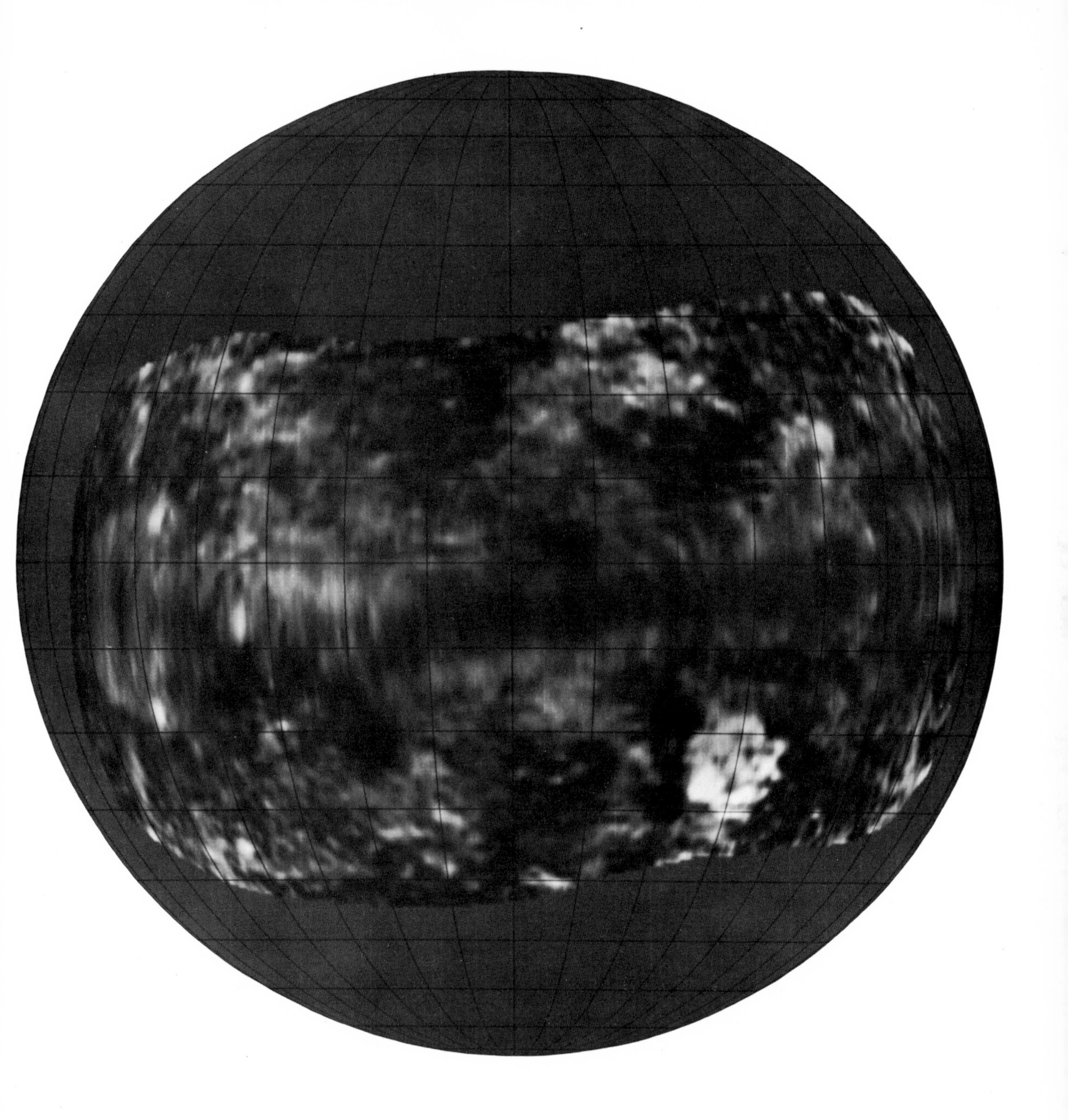

FIGURE 17. Radar map of Venus, made by Caltech's Jet Propulsion Laboratory. The map covers some 30 million square miles or about one-sixth of the planet's surface. The bright spot in the southern hemisphere is Alpha, believed to be a mountain range.

FIGURE 18. Venus thirty-four minutes before encounter on February 5, 1974. Launched from Earth on October 24, 1973, the spacecraft is now 19,600 miles from the planet.
FIGURE 19. Mercury eighteen minutes before encounter on March 30, 1974. Launched from Earth, October 24, 1973, the spacecraft is now 8,600 miles from the planet.

still further by Dr. Carl Sagan, who has suggested that even if there is no life on Venus at present, we might seed the atmosphere with suitable strains of bacteria. Over a few decades or centuries, they might break down the excessive quantities of carbon dioxide, release free oxygen, and allow the temperature to drop a few hundred degrees. And then there would be a new world for man . . . unless latter-day conservationists decide that the red-hot wilderness of Venus shall remain unspoiled.

With any luck, Mariner Venus-Mercury should increase manyfold our knowledge of this strange planet, and may help us to understand why our own apparently similar world has escaped its fate. The spaceprobe will observe Venus for about two weeks, before and after its closest approach (Figure 18). Very accurate navigation will be needed as it swings by the planet, for a one-mile error here will be magnified to a thousand miles by the time it reaches Mercury.

Although Venus has only one four-hundredth the mass of giant Jupiter, it will give the spacecraft a sizable boost as it flies past, swinging it in toward the Sun at the same time. Two months later—on March 30, 1974—it will pass Mercury (Figure 19).

Here, the TV cameras will have a perfect target. Mercury has virtually no atmosphere—perhaps as little as the Moon—so its surface features are never hidden. We have no idea what they are really like because the keenest-eyed astronomers with the best telescopes have never been able to see more than a few faint smudges on the face of the planet. It is impossible to observe properly from Earth, since at its closest it is between us and the Sun, so its dark face is then turned toward us.

Mariner Venus-Mercury should put an end to a century of eyestrain. During the two weeks while the spacecraft is flying past Mercury, it should radio back three thousand photos, showing details only a hundred yards wide.

After it has cut across the orbit of Mercury, MVM 73 will swing out past Venus, becoming

Chesley Bonestell

FIGURE 20. Venus and Earth in opposition, as seen from Mercury.

a little artificial planet with a "year" of 176 days. Now the year of Mercury is 88 days, or exactly half this. In other words, when Mercury has made two circuits of the Sun, MVM 73 will have made one—so *both will be back at the same place.* Thus with any luck, we may be able to arrange a second rendezvous with Mercury, 176 days after the first, and obtain another series of observations.

Little—3,000 miles in diameter—Mercury has recently been a source of much embarrassment to astronomers. They were completely confident that it always kept the same face turned toward the Sun, as the Moon does to the Earth—and for the same reason. The gravitational link between two close bodies can act rather like the brake blocks on an automobile wheel, slowing them down. The smaller one loses its rotation first, which is what has happened with the Moon and many other natural satellites.

The Solar System will not last long enough for tiny Mercury to check the Sun's spin, but everyone was quite certain that the other half of the operation was complete. A few astronomers, straining their eyes to glimpse details at the limits of vision, had even produced crude maps of the face which "forever" pointed toward the Sun. They always saw the same unchanging features, which appeared to settle the matter.

So arose an entire mythology of a planet where the Sun stood fixed in the sky, neither rising nor setting. On the dayside it would be so hot that lead or tin would melt; one could imagine metallic rivers meandering through a landscape of blistered rocks, with perhaps a volcano or two erupting in the background to give some local color. And on the nightside, in contrast, where the Sun had not shone for a billion years, it would be so cold that air would freeze solid. The sky would be black and ablaze with stars and planets—of which the most brilliant would be dazzling Venus, and the most fascinating the double planet Earth and Moon (Figure 20).

Between the eternal night and the eternal day would be a narrow twilight zone, a few hundred miles wide, where the Sun would be visible for about half the year. Perhaps this would be the best place for explorers to land, where they could have easy access to both hemispheres; the planet would obviously be of great scientific interest, but it was not anticipated that Mercury would ever be much of a tourist attraction.

In 1965, to everyone's amazement, this picture turned out to be quite wrong. Radar observations showed that Mercury had *not* become locked with one side toward the Sun; its day was not identical with its year of 88 Earth-days. It rotated on its axis in only 59 Earth-days.

How had the astronomers been fooled? It was quickly noticed that the numbers 59 and 88 were close to the ratio 2 to 3; Mercury had not achieved 1 : 1 synchronization, but it had done the next best thing, for reasons which are still very unclear. From the point of view of observers on the Earth, the results were much the same; the planet would always show the same features when it was best placed for viewing. No one had ever dreamed of a 2:3 ratio, and the simple 1:1 synchronism had been assumed. It had not even been realized that it *was* an assumption.

Einstein once remarked, apropos of the search for Nature's laws: "The good Lord is subtle, but He is never malicious." In the case of Mercury, however, He seems to have been having a small joke at the expense of terrestrial astronomers.

From the point of view of any most unlikely inhabitants (who, incidentally, should be called Hermians) Mercury's odd rotation has some very strange results. A 59-day spin does not mean that the sun will set and rise every 59 days, or anything like it. That would be true only if Mercury stood still on its orbit—and in 59 days it has gone two-thirds of the way around the Sun. In these circumstances (assuming a perfectly circular orbit) the Sun takes *exactly one (Mercury) year* to cross the planet's sky. The solar day of Mercury (dawn to dawn, or noon to noon) is therefore *twice* as long as the planet's year!

But—if you will kindly hold on for a little longer—there is one more fascinating complication. Mercury does not move in a circular orbit; its path around the sun is quite eccentric.

This means that it moves much faster at some times of the year than at others; as a result, the orbital motion of the planet can sometimes overtake the sunrise.

According to the Old Testament, the prophet Joshua commanded the sun to halt. Mercury can do even better than that. In some latitudes, at certain times of year, the Sun will rise, hover above the horizon—and set again. Then it will gather its strength, and rise once more to make its slow transit of the sky, smiting the land beneath with such fury that a Death Valley noon would seem scarcely warmer than the Antarctic winter.

What effect such extreme conditions can have upon a world, only a very rash geologist would care to predict—especially as Mercury has another peculiarity which makes it one of the most interesting bodies in the Solar System. It is much the densest of all the planets, indicating that it may contain a high percentage of heavy metals. Perhaps this is where most of the gold and platinum and uranium from the original solar nebula condensed, leaving only the tailings on the outer worlds. . . .

Even if this is true, it is hardly likely that transport costs would make Mercury Mines, Inc., a very attractive proposition. Yet one day, this strange little planet may be of great importance technologically, as well as scientifically.

The future of our civilization depends on power. We know that fossil fuels will be exhausted in a few more ticks of the clock, and uranium fission has come just in time to save us. Beyond that is the hope—rapidly approaching certainty—that the energy of fused hydrogen atoms can be controlled, and not merely released in the explosive violence of the H-bomb.

But even fusion power may not put off the day of reckoning more than a few centuries, if present trends continue. Whatever its source, we can only handle a limited amount of power on Earth before the waste heat which is the inevitable end of all thermodynamic processes becomes intolerable. "Thermal pollution" is already a local problem; early in the next millenium, it could be a catastrophic global one.

The only long-term solution is to go out into space, tap the Sun itself, and use its energies on the spot for whatever strange goals the technologies of the future may seek. What better place for doing this than Mercury, where the unshielded sun beats down from the sky with ten times the power we know on Earth, and underfoot are inexhaustible supplies of all the heavy metals?

But do not take these speculations too seriously. Mercury has already surprised us once; it may do so again.

SIX
THROUGH THE ASTEROIDS

The spacecraft that fly beyond Mars in the late 1970's will be basically very similar to the well-tested Mariners, and will use much the same hardware, control systems, cameras and sensors. But there will be one fundamental difference, which will be obvious even at a glance.

The outward-bound spacecraft will not be equipped with the sail-like solar panels carried by almost all earlier satellites and deep spaceprobes. They will be going so far from the sun that they can no longer collect enough of its weakening energy to provide the electrical power they need. Out at the orbit of Pluto—thirty-nine times Earth's distance from the Sun—the intensity of sunlight is less than one-thousandth of its value here, and some other source of power has to be found.

Fortunately, small nuclear generators, of the type already deployed on the Moon during the Apollo missions, provide a satisfactory solution. Fueled with plutonium 238, they can produce—*continuously*—several hundred watts of power for at least ten years, and as they contain no moving parts there is nothing to break down. The heat generated by the radioactive isotope is converted into electricity by a series of thermocouples—combinations of metals which, when heated at one junction and cooled at another, produce an electric current.

"Radioisotope thermoelectric generators" (RTG's) also have other important advantages for deep-space missions. The excess heat they produce is welcome, out around Pluto, to keep the spacecraft warm; it would cool off to about −370°F if it had to rely on sunlight alone. And as RTG's generate power continuously, there is no need for the batteries which solar-powered spacecraft must carry, to keep operating when they enter the shadow of a planet. (In the *shadow* of Pluto, incidentally, it would no longer be a balmy—370°F. The thermometer would descend almost to absolute zero, or about −450°F.)

With its usual fondness for acronyms, which must have caused a high mortality rate among lexicographers, NASA has coined the name TOPS—Thermoelectric Outer Planet Spacecraft—for this type of probe. A TOPS launch can take place, it is interesting to know, only with the written authority of the President—after he has had the assurance of the Atomic Energy Commission that several pounds of orbiting plutonium do not violate the United Nations Treaty on Outer Space.

Another striking difference between TOPS and earlier spacecraft is the size of the radio antenna carried. The Mariners used parabolic reflectors three feet across, which were adequate to beam signals back over distances of a hundred million miles or so. But *billions* of miles are involved in outer planet missions, so a 14-foot dish will be needed. Because of its size, it will have to be unfurled in space, like a sunshade. With only 40 watts of radio power, this dish will be able to send back a picture from Neptune—three billion miles away—every half-hour. And from Jupiter, a mere half-billion miles away, pictures can be received almost continuously, with no delay, except for the unavoidable forty-minute transmission lag due to the finite speed of radio waves.

The greatest problem involved in a spacecraft which must fly a mission lasting more than ten years is, of course, reliability. We are still a long way from attaining the standards set by the human brain (at least ten billion components, mean time to failure, seventy years or so) but the results already achieved are impressive. Mariner 4 contained 39,220 electronic parts; every one worked perfectly throughout the mission. There are communications satellites and space vehicles now in orbit that are still functioning six years after launch.

The outer-planet spacecraft will be several times more complex than Mariner 4, their missions will last twenty times longer, and they will fly through a much more hazardous environment. It would be absurd to expect 100 percent reliability, no matter how carefully all components are selected and assembled.

With TOPS, the only solution is to assume that something *will* go wrong, and provide a means of repair. The concept of redundancy—of "back-up" systems which take over in the event of failure—is an old one in space flight, but the TOPS designers have gone a good deal further. They have worked out a concept which has been christened STAR, for Self-Test And Repair computer. This electronic brain will have spares standing by for all its key elements, ready to be connected up instantly in case of failure. The decision to do this will be made by a Test And Repair Processor (TARP), which is continually checking all STAR's circuits.

But "who shall guard the guardian?" What happens if TARP itself has a brainstorm? To cope with such a crisis, the designers have fallen back on rule by committee. TARP is divided into three identical sections, like a brain with three lobes. If one lobe disagrees with the other two, it is outvoted, and if it does this too often, it is lobotomized and replaced with a spare. (I would like to think that it sings "Daisy, Daisy" as it is disconnected.)

By the use of such heroic and complex techniques, it is believed that STAR has at least a 90 percent—and perhaps a 99 percent—probability of success even on a ten-year mission. Of course, there is no way that the spacecraft could recover from a catastrophe like the explosion of a pressure tank, or a major mechanical failure. Only conservative design, and endless care in fabrication, can guard against these; it is not possible to duplicate *everything*.

Nor can there be any absolute protection against external dangers, such as meteorites and radiation. The statistics have to be examined carefully, and the risks accepted if they seem reasonable. This, after all, is the way every motorist operates when he leaves his garage—though luckily for his peace of mind he is seldom aware of it.

Before the opening of the space age, it was widely asserted that any outward-bound rocket would be riddled by a barrage of meteorites as soon as it left the protective shield of the atmosphere. Fortunately, this danger turned out to be grossly exaggerated; although spacecraft have been mildly pitted and sand-blasted by meteoric dust, there have been no serious incidents in thousands of years of total operating time.

A somewhat similar hazard confronts spacecraft on their way to the outer planets; it is hoped that this will also turn out to be overrated. It is the asteroid belt—a vaguely defined zone lying between Mars and Jupiter, and traversed by untold thousands of planetoids, or minor planets. The largest of these, Ceres, has a diameter of about 500 miles, but most are much smaller, and have been aptly described as flying mountains.

The discovery of the asteroids is a romantic little footnote to the history of astronomy.* Soon after the distances of the planets were established, it became apparent that there was a curiously large gap between Mars and Jupiter; the Solar System would look much tidier if it was occupied. So astronomers started to search for the missing planet. They thought they knew where to look because a simple mathematical law had been discovered (Bode's law) which gave the distances of all the known planets from the Sun with considerable accuracy. The law was purely empirical—even now, no reason has been found for it—but Mercury, Venus, Earth, Mars, Jupiter and Saturn obeyed it with remarkable accuracy *if* the existence of a planet X between Mars and Jupiter was assumed. When Uranus was discovered (Chapter 9), it too fitted Bode's law, and so the prospects for a new planet brightened considerably.

The philosopher Hegel ridiculed this numerology, and substituted some of his own. In 1801, he published a convincing proof of the *non*existence of any more planets; unfortunately for him, Ceres had already been discovered, on the first day of that year—January 1, 1801. Many years later the great mathematician Gauss remarked rather unkindly that though Hegel's thesis was "insanity," it was pure wisdom compared to his subsequent writings.

Yet Hegel did have a slight point: with a diameter of only 500 miles, Ceres was not much of a planet, and would barely serve as a good moon for Jupiter or Saturn. A little disappointed, the astronomers continued the search. During the next six years, they discovered three similar bodies—Pallas, Juno and Vesta, all between 300 and 100 miles in diameter. It began to look as if the Mars-Jupiter gap had once been occupied by a fair-sized planet, which for some reason had broken into pieces.

*For an excellent account, see Willy Ley's *Watchers of the Skies* (New York: Viking, 1963).

For almost forty years, no other planetoids were discovered, but thereafter, particularly with the advent of astronomical photography, they started to undergo a population explosion. By the beginning of the twentieth century, about five hundred were known, and the task of keeping track of them was getting quite out of hand. Before the era of electronic—or even desk—computers, it required several days of careful and tedious work to establish the orbit of a new planetoid, and until this was done, its discovery could not be authenticated. There are hundreds of examples of minor planets lost without trace, after one or two observations.

Today, about two thousand of these little worlds have had their orbits calculated with sufficient accuracy to establish their existence, but the astronomers long ago gave up naming them. Unless a planetoid has some exceptional characteristic, it is now given a number. By this time, all the larger specimens must have been discovered, and the ones turned up nowadays—usually while something else is being photographed—are mostly about a mile or so across. One cannot say "in diameter," since they are irregular masses of rock, probably very much like the little Martian moon Phobos (Figure 15). Only Ceres and a few other large plenetoids can have enough gravitational force to have pulled themselves into spherical shape; in powerful telescopes, it is just possible to see that they have circular disks.

The total number of asteroids has been estimated to be at least 50,000—but this depends upon where one stops counting. There may be a continuous gradation of size from Ceres down to objects no larger than grains of sand—and a grain of sand moving at 20,000 miles an hour, if it hit the wrong spot, could wreck a spacecraft. The number of particles larger than this must be enormous; there will be no way of counting them until we fly a mission through this region of space, and see what happens.

Certainly an attempt will be made to observe any convenient large asteroid that passes within range, and flight trajectories may be adjusted to allow this. If he relied on chance alone, an astronaut flying through the asteroid belt would be lucky to see even one of its

FIGURE 21. Left: Halley's Comet, May 15, 1910.
Right: the head of Halley's Comet, June 2, 1910.

inhabitants with the naked eye, but he could examine many through a good telescope—if he knew exactly where to point it. The orbital measurements patiently collected by astronomers for more than a century may soon have a value that they could hardly have imagined.

On occasion, a spaceprobe flying through the asteroid belt might also pass close to one of the most spectacular and mysterious of all the bodies orbiting the sun. Since the earliest times, comets have excited superstitious wonder, and were taken as omens of doom until Edmund Halley, in 1704, proved that they obeyed regular timetables. But even today we are very far from understanding their origin, and their behavior often defies all attempts to predict it.

The average comet moves around the Sun in a very elliptical orbit, which it may take thousands of years to complete. For more than 90 percent of that time it will be far beyond Pluto; indeed, it is possible that comets sometimes wander so far from the Sun that they are captured by other stars and never return. While it is traversing the outer part of its orbit, a comet usually appears in the telescope as a hazy, spherical blob of light with a tiny core or nucleus. But as it approaches the Sun, it undergoes an extraordinary transformation, becoming much more brilliant—so much so, in fact, that it may be visible *in broad daylight.* And it grows the characteristic tail (sometimes more than one) which may be scores of millions of miles long. A giant comet, in fact, is far larger than the sun, but it contains very little matter. Most of the impressive tail is gas so tenuous that it would be an extremely good vacuum by laboratory standards. Only the nucleus is solid, and probably consists of chunks of rock, meteoric iron and ice.

After a comet has rounded the sun, it departs into the outer darkness and sheds its tail. It may be back again in a few years—or in a few million years. The shortest known period is only three years; Halley's Comet takes seventy-five years to complete its orbit, so a lucky and long-lived man could see it twice.

A large comet would be a fascinating target for a spaceprobe, and though a rendezvous

would be difficult to arrange, a fly-*through* would be quite practical. An opportunity will arise in 1976 when Comet D'Arrest passes close to the Earth's orbit—but the really exciting challenge will be the next return of Halley's Comet, in February 1986 (Figure 21). However, there is a slight problem here. The orbit of Halley's Comet is tilted by 18 degrees to the plane in which the Earth moves—and, to make matters worse, it travels in the opposite direction. A spaceprobe launched toward it would meet the comet head-on at an enormous velocity—about 150,000 miles an hour. So the "fly-through" would be very brief, though it could still be scientifically rewarding.

Nevertheless, there is a possibility of a genuine rendezvous if we bring Jupiter into the act. A special type of spacecraft would be required, fitted with a low-thrust electric propulsion system which has already been tested out on a number of satellites. The spacecraft would be launched toward Jupiter in late 1977, and would arrive at the giant planet a year later. Jupiter's gravity would swing it around though more than 180 degrees—reversing its direction of motion—so that, like the comet, it would travel in a retrograde orbit. Then the electric propulsion system would be switched on, turning solar energy into thrust. By early 1985, the probe would match speed with the comet, and thereafter would cruise along with it forever, reporting back on its behavior as it swept around the sun and departed once more beyond the orbit of Pluto.

In May 1970 the distinguished astronomer R. A. Lyttleton delivered the prestigious Halley Lecture at Oxford, and ended with these words: "I venture to think that Halley would have been in favor of such a mission, and I will conclude by expressing the hope that when 1986 comes and brings Comet Halley back, the space fleets of all nations will be out to welcome it, to the greater glory of the illustrious Halley himself."

It is quite certain that there is some close relationship between comets, asteroids, and meteorites; they show many similarities, yet also many profound differences. There may be secrets hidden since the birth of the Solar System, waiting to be discovered in the no-man's-

PLATE I. Our own local galaxy, the Milky Way, as seen from a distance of 300,000 light-years. The white spot above the center nucleus marks the location of the Solar System. The galaxy contains at least 100 billion stars and is about 100,000 light-years across. In comparison, the Solar System is about 8 light-years across.

PLATE II. Mars from its outer moon, Deimos, 12,500 miles above the equator. The South Polar Cap is on the extreme right

PLATE III. A typical Martian landscape. The inner moon, Phobos, shows a barely visible disk.

PLATE IV. The zodiacal light as seen from Mercury, the foreground illuminated by Venus and Earth.

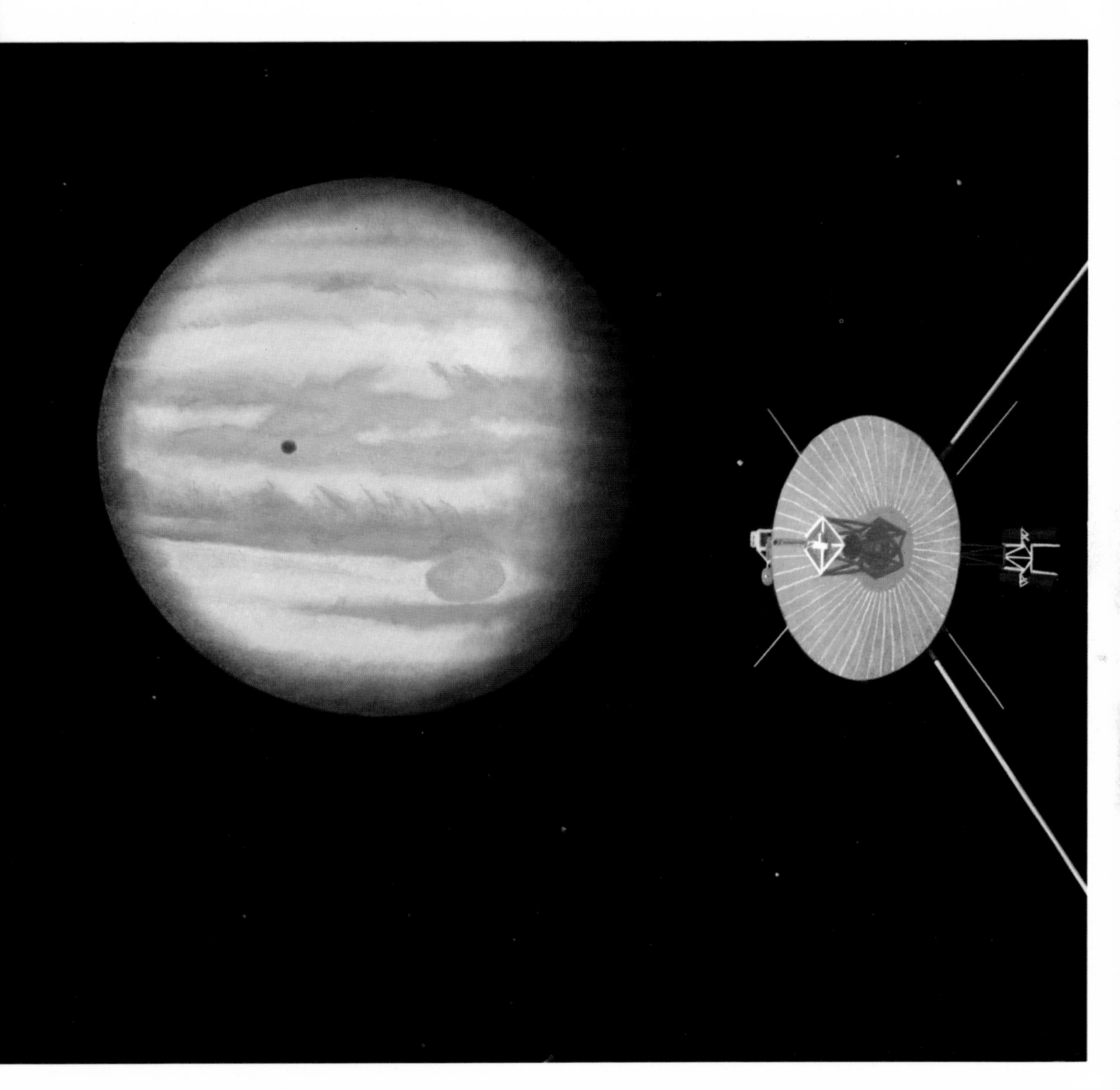

PLATE V. Jupiter four hours before encounter on March 1, 1979. The spacecraft was launched from Earth on September 4, 1977. Its distance from the planet is now 304,000 miles. The Great Red Spot is visible at lower right and the shadow of a satellite is in transit.

PLATE VI. The Solar System. Sun, planets and larger satellites are shown to scale; smaller satellites would actually be invisible. From left to right: Mercury, Venus, Earth, Mars, Jupiter, Saturn, Uranus, Neptune and Pluto.

PLATE VII. Jupiter from its closest large satellite, Io, 261,000 miles distant. The dark side of the planet can never be observed from Earth; auroral displays and chemical reactions in the atmosphere may permit some features to be seen.

PLATE VIII. Jupiter from its closest satellite, Jupiter V (Amalthea), 112,600 miles distant, showing typical cloud formations, the Great Red Spot and the shadow of an outer satellite in transit.

PLATES IX AND X. The two faces of Saturn's satellite Iapetus. Franklin and Cook have suggested that the brighter side is covered with "snow" and that the other side is nearly barren, the "snow"

having been eroded by a meteor swarm. In the "snow"-covered painting Iapetus is illuminated solely by light from Saturn; in the other, illumination comes from the Sun.

PLATE XI. Saturn from its largest moon, Titan, 760,000 miles away. Because of its size, (3,500 miles in diameter), Titan is one of the few satellites able to retain an atmosphere; hence the blue sky.

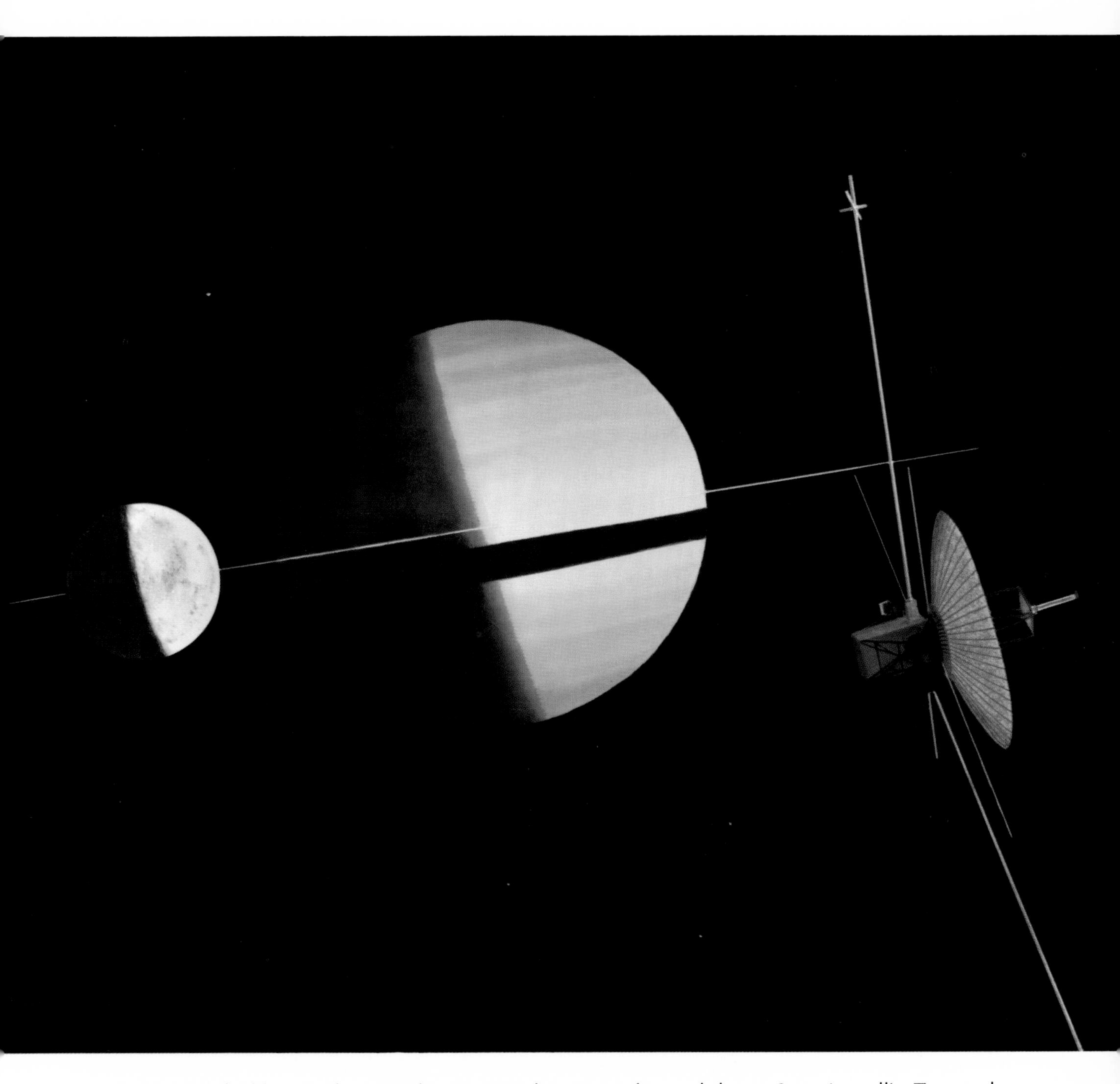

PLATE XII. Launched from Earth on October 17, 1977, the spacecraft passed close to Saturn's satellite Titan, and encountered Saturn on May 24, 1981. Now passing close to another satellite, Rhea, shown 15,000 miles away, the spacecraft will escape from the Solar System and join the stars.

PLATE XIII. The rings of Saturn as seen from just above the cloud layer at 15½ degrees north.

PLATE XIV. Uranus from its third moon, Umbriel, 166,000 miles distant. The second moon, Ariel, is also visible.

PLATE XV. Neptune from Triton, 220,000 miles away. Although appearing some 250 times the area of the full Moon as seen from Earth, Neptune would give only about the same amount of illumination because of its great distance from the Sun.

land between Mars and Jupiter. The asteroids may be remnants of a lost planet; or they may be fragments of one that was stillborn, because Jupiter's tidal pull prevented their aggregation into larger masses.

It would be of great interest to sample an asteroid, and perhaps we have already done so. The orbits of many meteorites lead back to the asteroid belt, so these lumps of rock and iron may be strays that have been deflected from their natural home by collisions, or the gravitational influence of Jupiter.

Quite recently it has been proved that a certain class of meteorite contains large amounts of complex organic compounds, including some that are the precursors of life. If these can form in an environment as unpromising as the asteroid belt, the probability of life in the universe may be very much higher than we have imagined.

Even a small, mile-wide asteroid has a mass to be measured in billions of tons; altogether, there may be about a tenth of the mass of earth (i.e., 1,000,000,000,000,000,000,000 tons) mostly of high-grade metals, circulating between Mars and Jupiter, and chopped up into pieces of convenient size. An imaginative space scientist, the late Dandridge Cole, calculated that a three-mile asteroid, if it had the same composition as the average meteorite, would contain about $50,000,000,000,000 worth of gold, platinum and other rare minerals. We may yet see grizzled prospectors working their way round the belt, with patched-up space shuttles taking the role of the traditional burro.

In his stimulating book *Islands in Space,* Cole discussed an even more startling idea: capturing an asteroid and bringing it back to Earth. In favorable circumstances, he calculated, a modest number of hydrogen bombs could kick an asteroid out of its orbit, and set it falling sunward. When it came near Earth, another explosion would be required to convert it into a second moon, which could be mined, or turned into a low-gravity health resort, or used as an appropriately neutral meeting place for the United Nations.

However, such an operation would be risky. If there was an error in navigation, the

FIGURE 22. The Great Siberian Explosion of June 1908. This painting shows the moment of impact, when what was probably a small comet hit the Earth. The explosion was so violent that the forest was devastated for sixty miles in all directions; 150 miles away men and horses were thrown to the ground; a jet of flame shot up twelve miles into the sky. In the center, scientists found more than two hundred craters, some of them fifty yards in diameter, and a shallow depression two miles across. The violence of the explosion was probably due to the instant expansion upon impact of the frozen gases that composed the comet's head. This, and the fact that no substantial amount of meteoric material could be seen, led scientists to the opinion that a comet rather than a meteor had struck.

asteroid might collide with Earth. There is a forty-mile-wide fossil crater in South Africa (the Vreedefort Structure) caused some millions of years ago by an asteroid only about a mile across.

Twice in this century (1908, 1947) much smaller objects have fallen in remote regions of the USSR and caused tremendous but, luckily, local damage (Figure 22). Although the asteroid belt may seem a long way off, what happens there could have a profound—even devastating—effect on human affairs. One day it may be a matter of life and death for millions, to deflect an Earth-bound asteroid from its course.

If our civilization lasts as long as we hope it will, this is not merely a remote and theoretical probability. It is as certain as tomorrow's dawn that, sooner or later, the Jovian bomb doors will open again.

SEVEN
THE WORLD OF THE GODS

Almost every year, for weeks at a time, Jupiter dominates the midnight sky. Yet it is not the most brilliant of the planets; Venus far outshines it, and Mars is often brighter. Despite this, it was Jupiter that the ancients honored with the name of their chief god. They chose better than they could have known, in the days before the telescope showed astronomers the real size of the planet.

From an explorer's point of view, the first important fact about any new and unknown territory is its area. In this respect, the challenge presented by Jupiter is staggering. Its surface area is *more than 120 times that of our own world.* Such a figure conveys very little meaning; the mind simply cannot grasp it. If some celestial big-game hunter peeled off the surface of our planet and pinned it upon Jupiter like a trophy, it would look about as large there as India does on a globe of the Earth. . . .

The mass of Jupiter is equally impressive—more than three hundred times Earth's. It outweighs *all* the other planets put together. With only slight exaggeration, Isaac Asimov once described the Solar System as "Jupiter—plus debris."

Yet despite its size, Jupiter spins much more rapidly than Earth; its "day" is just under ten hours. Thus whereas a point on our equator moves at 1,000 miles an hour, on Jupiter's it travels at 28,000 miles an hour—more than the velocity of escape from our planet. It is not surprising, therefore, that centrifugal force has given Jupiter a distinct bulge.

Even in a small telescope, Jupiter is a striking object, usually showing several bands of cloud running parallel to the equator—another sign of its swift rotation. These clouds are all that we have ever seen of the planet; the solid surface, if there is one, lies hidden hundreds or perhaps thousands of miles down at the bottom of an atmosphere which is very largely hydrogen.

As hydrogen is the lightest of all gases, nothing can float in it; and as it is also quite transparent, we should be able to see through the Jovian atmosphere. The composition of the clouds therefore presents some major puzzles. Although the spectroscope reveals the presence of the hydrogen-rich compounds ammonia (NH_3) and methane (CH_4), these gases are colorless and cannot account for the delicate pinks and salmons that make Jupiter such a beautiful object (Plate V).

What's going on down there? Certainly a great deal of high-pressure chemistry. Experiments with mixtures of gases imitating the Jovian atmosphere have shown that, in the presence of radiations and electric sparks, complicated soups of organic compounds are formed. Many of these chemicals are the building blocks of life itself, and they also exhibit some of the colors observed on Jupiter.

Now this is quite exciting, and has caused a revolution in our ideas about the planet. Until a few years ago, the concept of Jovian life would have been dismissed as absurd by almost all scientists. Not only is a hydrogen-methane-ammonia atmosphere completely poisonous, but Jupiter's great distance from the Sun (five times that of Earth), indicates that it must be extremely cold. The calculated temperature turns out to be a chilly *three* hundred degrees (Fahrenheit) below freezing. . . .

But the calculations have proved to be wrong; measurements made from jets flying at great altitude to avoid the effects of the Earth's atmosphere show that Jupiter is much hotter than expected. It does not, therefore, depend only upon the Sun for warmth, but must have some internal source of heat. If we went down into the atmosphere, we would eventually reach a level of comfortable temperature—though probably not of comfortable pressure.

As for the "poisonous" nature of the atmosphere, that merely depends upon the point of view. We now feel fairly certain that the primitive Earth had an atmosphere completely lacking in oxygen and, like Jupiter's, rich in methane and ammonia. Here were synthesized, by the action of sunlight and other energy sources, the hydrocarbons and amino acids upon which life depends.

This process took only a very short time; it happens in a few hours in the laboratory. After that, it may have required millions of years for the first primitive, single-celled plantlike organisms to arise. They promptly started to pollute the environment with their waste products —notably oxygen. But in poisoning themselves, they created an opportunity for the higher life-forms represented by the animal kingdom. It might be argued that we are now busily preparing for the next step: making a world fit only for machines.

Be that as it may, the "chemical evolution" which preceded life may now be progressing on the grandest possible scale beneath the clouds of Jupiter. And if this is true, it would be unreasonable to suppose that the process has stopped there. Jupiter may be teeming with primitive life-forms—and perhaps some of them are not so primitive. Dr. Carl Sagan, one of the most eloquent advocates of Jovian biology, has speculated about creatures which could float at levels where the atmosphere was dense enough to support them, browsing like aerial whales on the snowfall of organic compounds.

Even if Jupiter turns out to be as lifeless as the Moon, it will still present enough problems and puzzles to keep scientists busy for centuries. One of its most famous enigmas is the Great Red Spot (Plate VIII), an oval-shaped formation several times the size of Earth. The Spot has been studied for more than a hundred years, so it is (at least by human standards) a permanent feature. But it is not fixed; it slowly drifts against the general background of clouds. It also shows major changes in color, and on occasion disappears completely for years at a time.

Theories to explain the Spot are about as numerous as the astronomers who have studied it. In some ways, it appears to behave like a floating island, but even if Jupiter has liquid oceans far down beneath the atmosphere, it is hard to understand how anything can rear up out of them, so high as to be visible from space. The giant planet has a giant gravity—two and a half times that of Earth. A Jovian Everest could be only about two miles high.

The most widely accepted theory today is that the Great Red Spot is some type of atmospheric disturbance, a kind of permanent cyclone perhaps associated with an irregularity in the hidden surface far below. This could be a mountain or a crater. From time to time, Jupiter must certainly be impacted by asteroids caught in its gravitational field. Perhaps the Great Red Spot is the slowly fading monument to a lost moon.

The most recent of Jupiter's surprises is the discovery that it emits powerful blasts of radio waves, in the ten-meter band, from a few localized spots deep in the atmosphere. This sounds startlingly like science fiction, but the waves show no sign of intelligent modulation—they are pure noise, resembling the "static" produced by thunderstorms. The energy involved is enormous, sometimes equivalent to an H-bomb *every second,* and there is still no satisfactory theory to account for these "noise storms."

At much shorter wavelengths (tenths of a meter) Jupiter broadcasts steadily from an invisible halo extending out into space for a distance several times the diameter of the planet. This discovery was the first proof that Jupiter has an equivalent of Earth's Van Allen zone—immense, and potentially dangerous, clouds of charged particles trapped in its magnetic field.

Jupiter has more moons circling it than the Sun has planets—and four of these twelve satellites are full-sized worlds. Probably there are many others still awaiting detection, but they cannot be more than a few miles in diameter.

The four giant satellites—Io, Europa, Ganymede, and Callisto—range in diameter from 2,000 to 3,500 miles. They are easily visible in a small telescope, or even a good pair of binoculars, and their changing positions from night to night have fascinated observers ever since they were first seen by Galileo in 1609. They form an extremely regular system since they move around Jupiter in almost perfect circles, precisely above the equator.

Much closer to the planet than any of the four Galilean satellites is a peculiar little moon called Jupiter V; it has never been formally christened, though the name Amalthea is sometimes applied to it. J. V. is so close to Jupiter (Plate VIII) that it moves through the heart of the great radiation belt, and it races round the planet in a mere twelve hours—only two hours longer than the brief Jovian day. As a result, it drifts very slowly across the sky; if it was just a little closer to Jupiter, it would hover motionless above one spot on the planet, like one of COMSAT's synchronous communications satellites.

The remaining seven moons are all very tiny (only a few miles across) and revolve round Jupiter in wildly eccentric orbits at vast distances. Almost certainly they are stray asteroids which have wandered into the planet's gravitational field and have become trapped like flotsam in a whirlpool.

Practically nothing is known about the dozen worlds of this mini solar system; only the four largest satellites show visible disks, even in the most powerful telescopes. A few faint and quite meaningless markings have been observed on them—just enough to prove, as would be expected, that they always keep the same face toward their giant master, as the Moon does to the Earth. They must be very cold, and perhaps speckled with ammonia or methane frost; we can expect to learn nothing about their surface details until we obtain close-ups from TV-equipped probes.

The first such probe, Pioneer 10, was launched in the early spring of 1972 and is targeted for arrival at Jupiter in December 1973; its successor is due to be launched at the next opportunity—which, as explained on page 11, is thirteen months later. These will be the first of all man's spacecraft to pass through the asteroid belt; indeed, after their encounters with Jupiter, they will have gained enough energy to leave the Solar System. Who could have dared to predict that, only a dozen years after we learned to escape from the Earth, we would be able to escape from the Sun!

The aptly named Pioneers will carry instruments which, during their two years of flight, will radio back information about magnetic fields, meteorites, and radiation and solar phenomena in regions of space never before explored. When they reach Jupiter their two-color cameras will take the first close-ups of the planet—and, it is hoped, of its satellites as well.

It would not be difficult to modify this type of spacecraft to go into orbit round Jupiter—

like Mariner 9 around Mars — or to send probes down into the planet's atmosphere. The probes would enter at an enormous velocity — over 130,000 miles an hour! — and would be destroyed by friction; but even if they survived no more than a few seconds, they could radio back information obtainable in no other way.

Much later, though perhaps during this century, it might be possible to establish floating robot observatories, or buoyant probes, at various altitudes in the Jovian atmosphere. From these, the levels far below could be investigated by sonic, radio and electrical instruments. And then the true picture of this extraordinary and enormous world would slowly come into focus.

Perhaps the closest analogy we have on Earth to the exploration of Jupiter is the investigation of the ocean deeps. Only recently have we developed techniques for doing this; yet we have been sailing the seas for at least three thousand years. The pressures in the Jovian atmosphere may be far greater than at the bottom of the Marianas Trench.

Whatever surprises we discover on Jupiter — and some of them may be shattering, especially if the wilder hopes of the biologists are true — the exploration of this huge planet may have a profound psychological effect on the human race. The average man has not yet waked up to the fact that next-door-but-one there is a world ten times the size of Earth.

Our very first Pioneers will reach it in less time than Magellan's ships took to circumnavigate the globe, four and a half centuries ago; now, any jet traveler can make *that* journey in two days.

Perhaps we will never get to Jupiter in two days. But we will do it in two weeks — much less than four and a half centuries from now.

EIGHT
LORD OF THE RINGS

The initial reaction of many people, when they first see the planet Saturn hanging in the field of a medium-powered telescope, is one of sheer disbelief. Such an extraordinary object scarcely seems real; it looks almost artificial, like the discarded toy of some infant god. One can well imagine the astonishment of Galileo and his successors when they caught glimpses, through their crude instruments, of a planet which appeared to have handles.

Saturn is Jupiter's smaller brother, with a diameter about 20 percent less, yet it has only a third of Jupiter's mass. In fact, the average density of Saturn is less than that of water; in a large enough sea, the planet would float like a piece of wood.

Physically, Saturn must be similar to Jupiter, with a very deep hydrogen atmosphere surrounding a solid core at an unknown depth. And like Jupiter, it has a ten-hour day; the equatorial bulge caused by this rapid rotation is very pronounced. In the telescope, the planet appears a pale yellow with faint dusky bands parallel to the equator. White spots and other markings sometimes appear, but atmospheric disturbances are much less common than on Jupiter; Saturn is a cooler and quieter world.

But it is certainly a spectacular one, thanks to its beautiful rings (Plates IX–XIII). Only about a mile thick, yet 170,000 miles across, they provide the closest approximation in the known universe to a perfectly flat, two-dimensional object. To make an accurate model of the rings, using the thinnest airmail paper, it would be necessary to cut out a circle thirty feet across. Making it lie absolutely flat and unwrinkled would be quite a feat.

Yet Saturn manages it. For more than a century it has been known that the rings are not solid; they consist of myriads of particles, or micro-moons, moving round the planet in independent, perfectly circular orbits. If they were not perfectly circular, there would be continuous collisions and the ring system would slowly disintegrate. Perhaps this is happening; we may be lucky to live at a time when we can enjoy such a beautiful, but ephemeral, phenomenon.*

*Possibly Jupiter also had a ring at one time. Some astronomers have looked for traces of it, without conclusive results.

The gaps between the three main rings are caused by Saturn's closer moons; they produce tidal or gravitational effects (resonances) which make certain positions unstable; any particles that wander into these forbidden zones are quickly ejected.

If any proof was needed that the rings cannot be solid, it is provided by the fact that quite faint stars can be seen shining through them. Exactly what the rings would look like at close quarters is still very uncertain; they may consist of mountain-sized objects many miles apart, or a much larger number of small rocks and boulders, probably covered with a film of ice. This question may be settled during the Grand Tour, but from a respectful distance. It will be a long time before any spacecraft actually enters the rings, though once one matches speed with them, it should be perfectly safe there. In any given region, all the particles would be moving with identical velocities, so an observer would appear to be in a cloud of motionless rocks or small icebergs. From time to time there might be an impact, but it would be so gentle that it could do no harm.

Like Jupiter, Saturn has an impressive and far-flung family of satellites—ten in all. Titan, the largest, is 3,000 miles across, or about the size of the planet Mercury; the other moons range from about a thousand down to a hundred miles in diameter. Whereas most of Jupiter's satellites have never been christened, all of Saturn's have glamorous names taken from Greco-Roman mythology. Listed in order of their distance from the planet, they sound like a line of Miltonian verse: Janus, Mimas, Enceladus, Tethys, Dione, Rhea, Titan, Hyperion, Iapetus, Phoebe.

Because of their great distance from Earth—never less than 800,000,000 miles, or some three *thousand* times the distance from Earth to our Moon—it is not surprising that we know practically nothing about them. Even the few facts that have been fairly well established are rather puzzling. Mimas and Tethys, for example, appear to have such a low density that they cannot consist of rock; it has been suggested that they are something like giant snowballs. But Tethys is at least 500 miles in diameter, and it is hard to imagine a snowball *that* size.

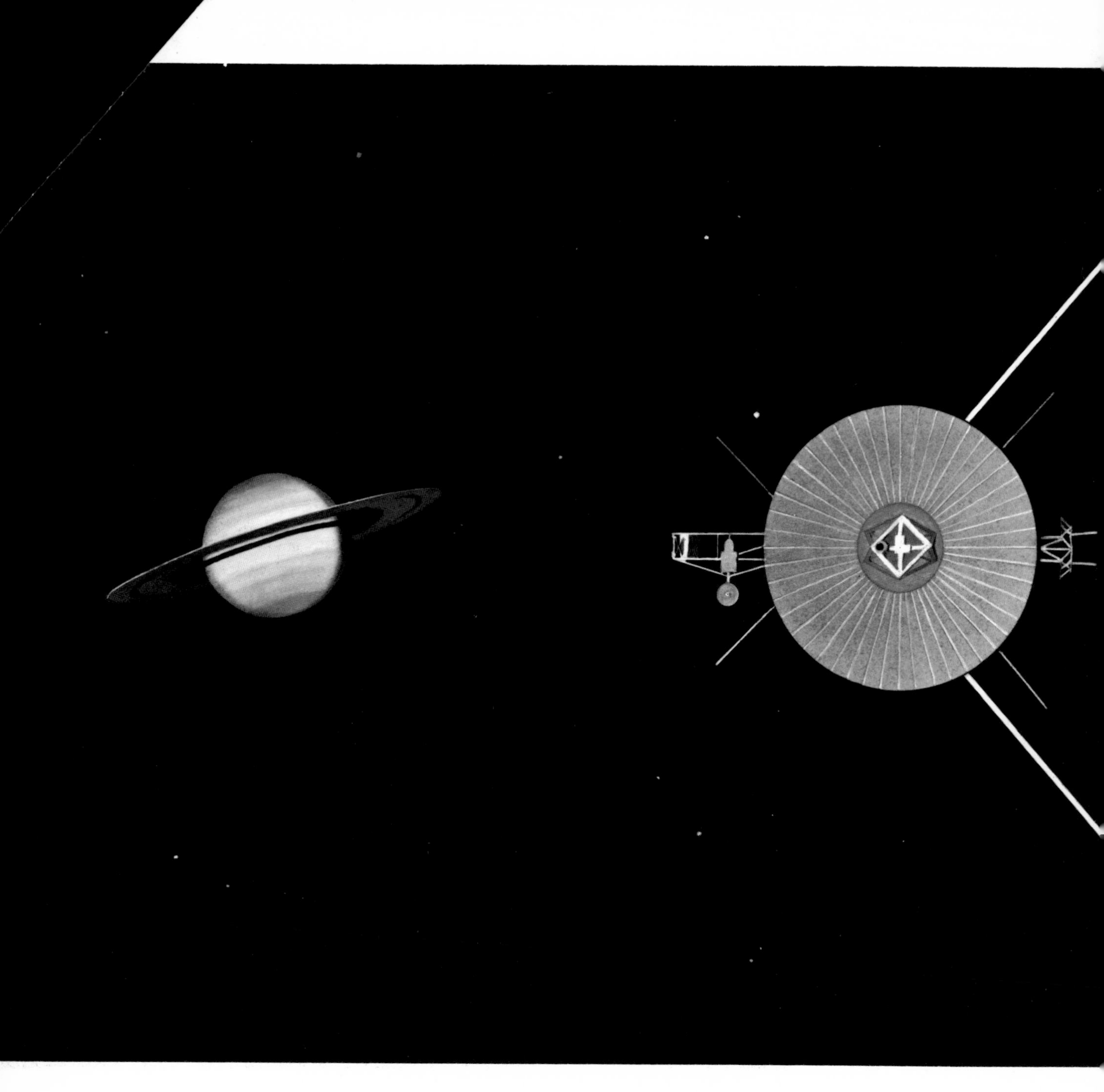

FIGURE 23. Twenty-four hours before encounter with Saturn on July 11, 1980. The spacecraft is still 1,300,000 miles from the planet; its radio dish is pointed directly toward Earth, which it left on September 4, 1977.

FIGURE 24. At the moment of encounter with Saturn on November 7, 1980. The spacecraft was launched from Earth on September 4, 1977, and is now passing over the planet's southern hemisphere at a distance of 338,000 miles. After deflection by Saturn, it will reach Pluto on March 9, 1981.

FIGURE 25. (OVERLEAF) Saturn from its third moon, Enceladus, 148,000 miles out. The second moon, Mimas, is visible in the plane of the rings.

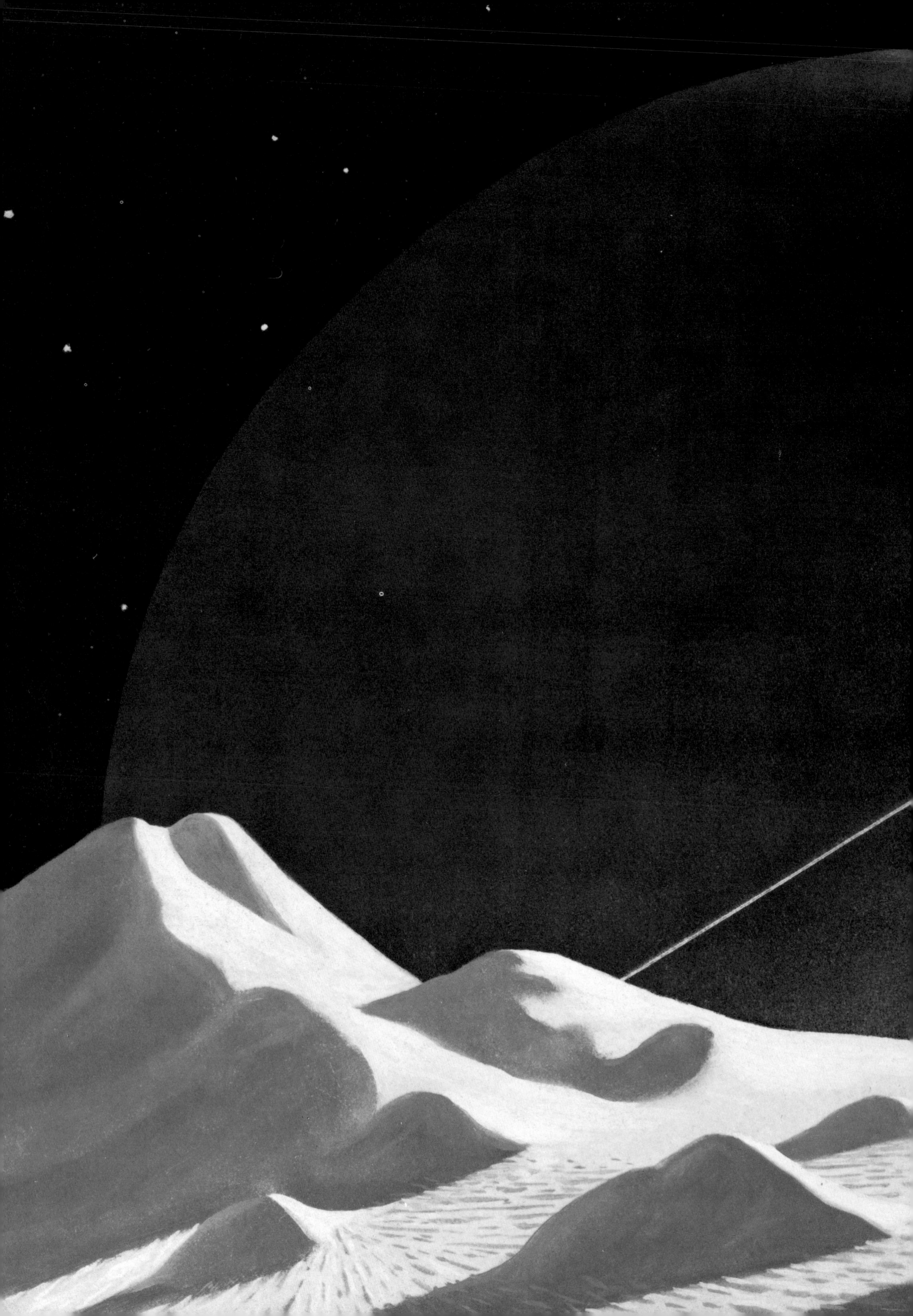

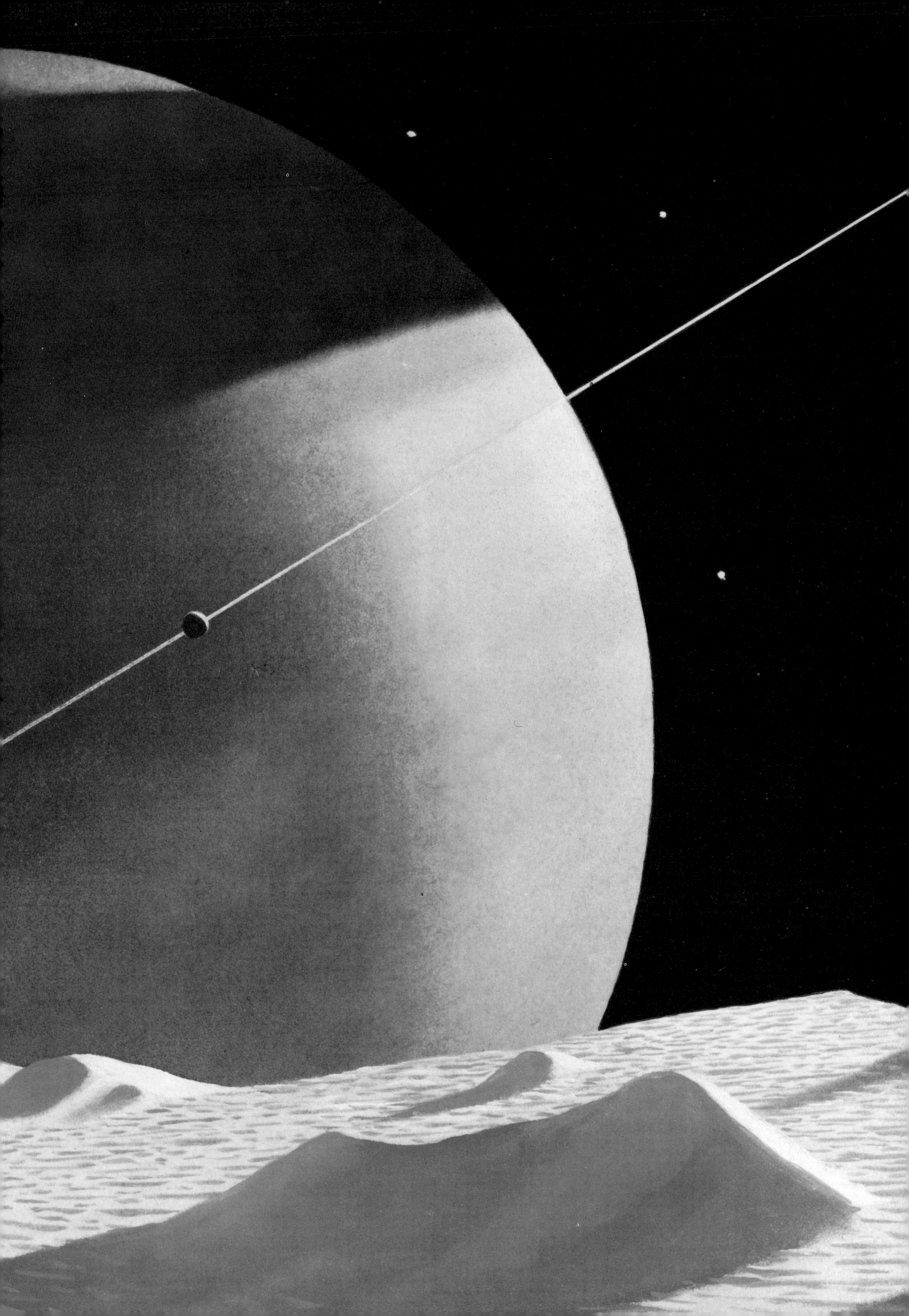

Either the observations are wildly inaccurate (which is quite possible) or something very odd is going on there.

In the case of Iapetus (or Japetus), a satellite with a 1,000-mile diameter, there can be no doubt that something odd *is* going on. Three hundred years ago, the astronomer Huygens noted that Iapetus was far brighter at the western end of its orbit than at the eastern. The variation in brilliance is astonishing—six to one; it is almost as if Iapetus were flashing on and off. Presumably there is a highly reflecting area on one side, though why Iapetus has been selected for this special treatment is a major mystery.* One possible explanation is shown in Plates IX and X.

Saturn's largest moon, Titan, also has a claim to distinction; it was the first satellite found to have an atmosphere—one of methane. Probably most of the larger moons of Jupiter and Saturn have similar atmospheres, though they may be too thin to be detected except by instruments on the spot. Methane remains a gas even at temperatures down to −260°F, long after ammonia, carbon dioxide, and most other gaseous compounds have condensed.

Although the view of Saturn from its various moons, especially a close one like Enceladus, is spectacular, the rings cannot be seen to the best advantage from the inner satellites. As Figure 25 shows, they are then almost edge-on and appear as a thin line of light bisecting the planet—and sometimes casting a broad shadow across it. There will probably be complaints about this one day from visiting tourists, misled by the posters put out by the satellite hotels.

From Earth, we can get a more open, though of course much more distant, view of the rings. Every fifteen years they are edge-on (and almost disappear, except in large telescopes),

*In the novel *2001: A Space Odyssey*, I placed the "Star Gate" monolith on Iapetus, at the center of "a brilliant white oval, about four hundred miles long and two hundred wide." The astronomers A. F. Cook and F. A. Franklin came to a similar conclusion in their study of the satellite's brightness variations, and report with a straight face: "Our picture of the current state of Iapetus has, to some extent, been anticipated on rather different grounds by Clarke. We are, however, unable to substantiate his claim that the center of the bright region contains a curious dark structure" ("An Explanation of the Light Curve of Iapetus," *Icarus*, Vol. 13, 1970).

but then they open up, and under the most favorable conditions appear very much as in Plate IX.

However, only a spaceprobe flying near the pole of the planet will see the rings really wide open (Figure 24). This type of encounter will be necessary in a mission to Pluto, as explained in Chapter 2. So one valuable and unexpected bonus from the exploration of the most distant planet will be a new look at the most beautiful one.

Well, the second most beautiful one. . . .

NINE
THE LAST OF THE GIANTS

Almost a billion miles starward from ringed Saturn lies another gas giant, Uranus, and a billion beyond that, a fourth one, Neptune. In size, they are very nearly identical twins, their diameters being about 31,000 miles, or four times that of Earth. Because of their great distance and poor illumination—out at Neptune, the sunlight has dropped to barely a thousandth of its value on Earth—they normally appear in even the largest telescopes as small, featureless disks, greenish in hue. Occasionally, Uranus shows faint bands (Plate XIV), similar to those of Saturn, but no markings have ever been observed on Neptune.

Their low density (like Jupiter's, about one and a half times that of water) indicates that they too are largely composed of hydrogen. The spectroscope has detected vast quantities of methane gas; ammonia is also probably present, but it will have frozen out into crystalline clouds.

Uranus has one peculiarity which must give it a very strange climate, if one can apply that word to a planet where the thermometer may never rise to −300°F. Its axis is tilted through no less than 98 degrees; in other words, its poles lie almost in the plane of its orbit. For much of the time, therefore, the planet appears to be rolling around the sun. (Our own Earth has its axis almost at right angles to the plane in which it moves; nevertheless, that modest 23-degree tilt produces all the seasonal changes, from summer heat to winter cold.)

The "year" of Uranus is longer than most human lifetimes—84 Earth-years. Its seasons must therefore average 21 years in length, and for most of one "summer" the north pole will be lying directly beneath the Sun. Forty-two years later, the situation is reversed, and the south pole becomes tropical—by the modest standards of Uranus. One would expect this annual switch to produce violent changes of atmospheric circulation, and the meteorology of Uranus may turn out to be unimaginably strange.

No one knows why Uranus has flipped over on its side. What is equally remarkable, it has taken all its satellites with it. There are five of them, poetically named (in order of distance) Miranda, Ariel, Umbriel, Titania, and Oberon, and they revolve almost exactly in the plane of Uranus' equator. Perhaps it acquired them *after* it toppled. All of them are quite small, ranging between 200 and 600 miles in diameter. Although they are close enough to Uranus to show visible disks, they must be very dim and inconspicuous objects in its sky.

Uranus has the distinction of being the first planet found in historic times. All the others, from Mercury out to Saturn, have been known from antiquity, and numerologically minded philosophers had gone on record "proving" that there could be no more than the classical five. But in 1781 William Herschel created a sensation when, with a six-inch reflector he had built himself, he spotted an object just a little too fuzzy to be a star. It turned out to be Uranus. Fortunately, Herschel's patriotic attempt to name it Georgium Sidus after His Majesty King George III was not well received, particularly in the United States. Under very favorable conditions, Uranus can just be seen with the naked eye, so in theory it might have been discovered at any time since men started making careful observations of the skies.

That is certainly not true of Neptune, whose discovery was even more romantic—and controversial—than that of Uranus. After the astronomers had been watching their new planet for several decades, they were surprised to find that it was not following the orbit they had calculated; some unknown force was disturbing it. Two brilliant mathematicians—J. C. Adams, an Englishman, and U. J. J. Leverrier, a Frenchman—tackled the problem. They assumed that an unknown planet might be producing these disturbances through its gravitational attraction, and set out to find it—arriving at virtually the same answers almost simultaneously. Leverrier's figures led to the actual discovery in 1846, and as might be expected, it triggered an unseemly Anglo-French squabble over priority.* It later turned out that an astronomer at the Paris Observatory had recorded Neptune's position on two occasions fifty years earlier. Assuming that he had made a mistake because the "star" had moved in the two days since he first charted it, he threw away his observations—and his chance of immortality.

It has been said, rather unkindly, that by far the most interesting thing about Neptune is

*But not between Leverrier and Adams, who remained lifelong friends.

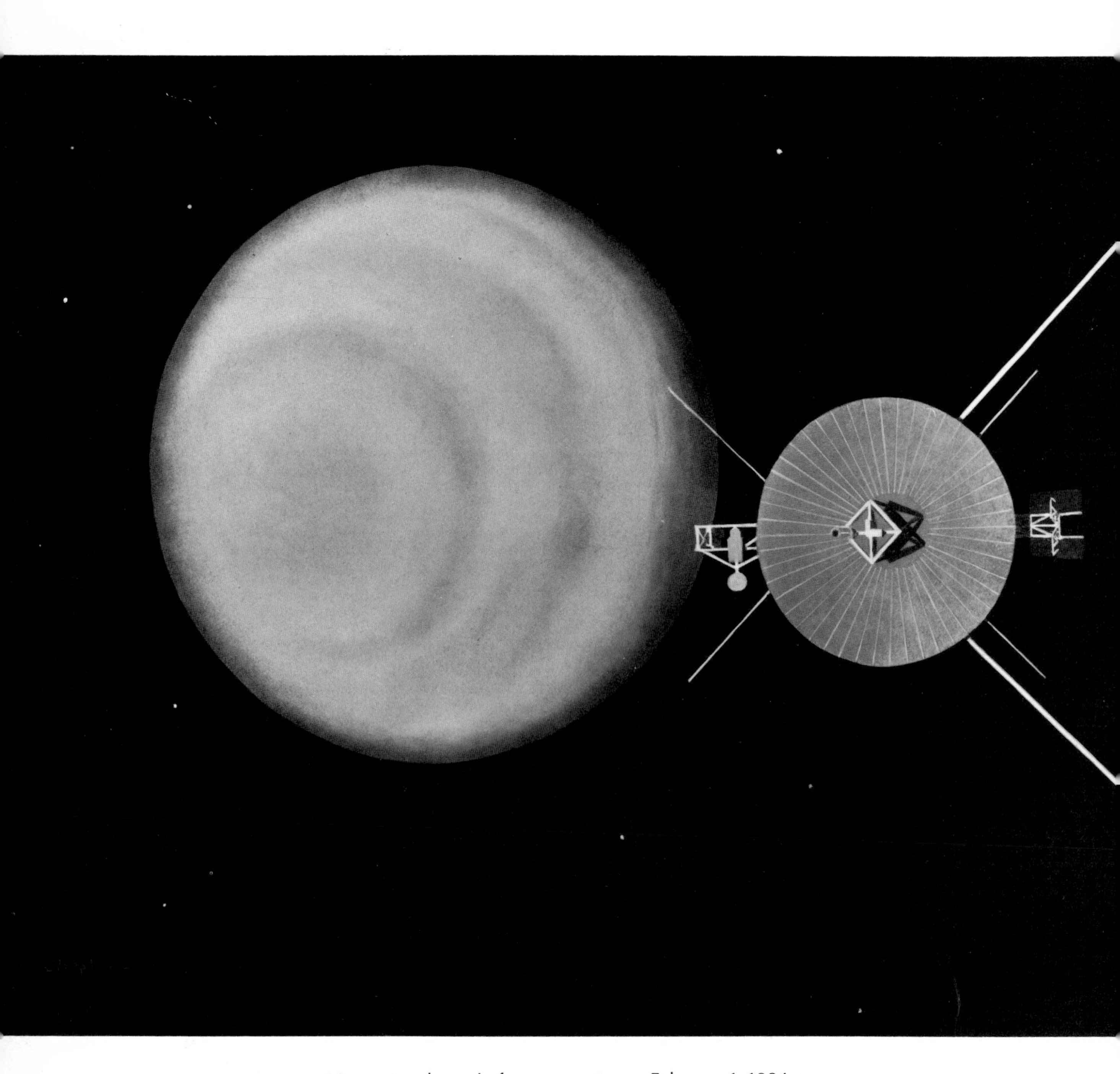

FIGURE 26. Uranus two hours before encounter on February 1, 1984 (launch from Earth January 29, 1979). The spacecraft is 90,000 miles from the planet. Because the axis of Uranus lies almost in the plane of its orbit, the spacecraft comes in over the pole.

FIGURE 27. After the encounter with Pluto on March 9, 1986 (launch from Earth, September 4, 1977). The Sun is too small to show a visible disk and its light is one thousand times weaker than on Earth.

the story of its discovery. This will certainly change, for there must be a great deal to learn about a planet with four times the diameter and seventeen times the mass of our entire Earth. Like the inner giants, Neptune has a deep hydrogen-methane atmosphere, and must be exceedingly cold — about 400°F below freezing, unless it has internal heat sources.

For a long time it was believed that Neptune had only one satellite, a large moon (3,000 miles in diameter) appropriately named Triton. Strangely enough, although Triton moves along a nice circular orbit with no measurable eccentricity, it travels *backward* — that is, against the direction of Neptune's own sixteen-hour spin.

Triton was discovered within a few weeks of Neptune itself; a second and much smaller satellite was found just over a hundred years later, in 1949. Called Nereid, it has the most eccentric orbit of any known moon. At its closest, it is only 832,000 miles from Neptune; at its furthest, it is more than 6 million miles out. But unlike Triton, it does at least revolve around Neptune in the right direction — taking nearly one Earth-year to do so. (Neptune's own year, incidentally, is 165 times longer than ours. We have not yet seen it make one circuit of the Sun; for that, we will have to wait until the year 2011.)

The prediction and subsequent discovery of Neptune led to great rejoicing among the astronomers, who now felt that at last they really understood the workings of the Solar System. Their satisfaction was premature; further studies showed that the orbit of Uranus still had unexplained irregularities. So more calculations were made, and a search was started for yet another planet. It led, in 1930, to the discovery of Pluto by Clyde Tombaugh at the Lowell Observatory, Flagstaff, Arizona.

This small, dim world now marks the known frontier of the Solar System; only comets wander beyond it. But Pluto has turned out to be a disappointment — and a puzzle. It is far too small to account for the observed perturbations of Uranus and Neptune; the calculations indicated a large body, and Pluto is only half the size of Earth. Perhaps another and more massive planet is still lurking out there in the darkness; if so, it will be very hard to find.* Dr. Clyde Tombaugh, during and after the quest for Pluto, spent a total of seven thousand hours examining an unbelievable 90 million star images†—but he never repeated his initial success.

Pluto is almost identical with Mars in size, mass and density; it should clearly be classed among the inner "terrestrial" planets, not among the outer gas giants. Not only does it seem to be in the wrong place, but everything else about it is anomalous. Its 250-year orbit is the most highly inclined, and most eccentric, of any planet. As a result, it can approach closer to the sun than Neptune does. In fact, for the twenty years from 1979 to 1999 Neptune, not Pluto, will be the outermost planet; because of the orbital inclinations, however, there is no possibility of collision.

These peculiarities have suggested to some astronomers that Pluto is really a lost moon—perhaps an escapee from Neptune—rather than a full-fledged planet in its own right. Some day, when we have really accurate information about the masses and movements of all the planets, our computers will be able to run the Solar System backward and see what happened to it in the remote past; then we will know if this theory is correct.

The basic knowledge required before this can be done may only be obtainable from spaceprobes. One of the most important by-products of the Ranger, Orbiter and Mariner flights was exceedingly accurate information about the orbits and masses of the Moon, Venus and Mars. Because a spacecraft is continually emitting radio waves of precisely known length, the tracking stations on Earth can locate it with extreme accuracy. To use a rather far-fetched analogy, it is almost as if the probe were continually playing out an invisible

*In the spring of 1972, Dr. Joseph L. Brady of the Lawrence Livermore Laboratory reported that his calculations—based on the anomalous behavior of Halley's Comet—indicated a giant planet beyond Pluto. It would be about three times the mass of Saturn and have a period of 464 years.

†He still has millions of unexamined star images. If there are any volunteers . . .

tape measure behind it, marked off in wavelengths. This accuracy would have been quite unbelievable before the electronic age. At the orbit of Mars, 40 million miles out, the positions of the Mariners were known to within a few yards.

So when the spacecraft enters the gravitational field of a planet, we know exactly what is happening to it. From this, we can calculate backward and discover not only the mass of the planet, but even its shape and the manner in which its density may vary from point to point. This was the way in which the lunar "mascons" (mass concentrations causing local highs in the Moon's gravity) were discovered; the same principle can be applied to all the planets.

It has often been said that when we explore space, we are also exploring time. Only when we have flown missions to every part of the Solar System will we have the vital statistics of all its components. And only then will we be able to turn back the pages of the book of cosmology—to the origins of our own world, and perhaps of the universe itself.

TEN
INTO THE ABYSS

Seven or eight years after it has been boosted by the slingshot of Jupiter's gravitational field, a Grand Tour spacecraft will reach the known limits of the Solar System. It will still be moving under the gravitational influence of the Sun, very slowly losing speed—but the Sun will never be able to call it back, for it will have far exceeded "solar escape velocity." When the Sun has done its utmost to slow it down, the spacecraft will still have a residual velocity of at least 50,000 miles per hour. At this speed, it will head out across the interstellar gulf—toward the stars.

The nearest of the stars, Proxima Centauri, is 4.3 light-years, or 25,000,000,000,000 miles away. If the spacecraft was aimed in that direction, it would get there in about fifty thousand years. But by that time, of course, Proxima would be somewhere else . . . because all the stars are being swept along in the giant cosmic whirlpool of the Milky Way, which turns once in every 200 million years.

At this primitive stage of our technology, therefore, interstellar probes are hardly practical. Even if we were patient enough to wait until they arrived at their destinations, they could not carry powerful enough transmitters to send back useful information from even a small fraction of a light-year away. And it is hard enough to build complex electronic equipment which will function reliably for a decade—let alone for thousands of years.

Yet the idea is not fundamentally absurd; what has already been done in the first generation of spaceprobing would have seemed utterly impossible not long ago. Perhaps in a hundred years, at the present rate of progress, we may be able to build vehicles that can reach an appreciable fraction of the speed of light. In theory, nuclear energy is quite capable of giving us this performance—and sooner or later, practice usually catches up with theory.

Such probes could reach the nearer stars in a few centuries, and a stable world-society might not think such a time scale unreasonable. (Remember the cathedral builders of the Middle Ages, planning generations ahead, in eras of plagues and wars.) And the energies that launched these far-ranging travelers would also be able to power their radio transmitters, so that they could send back to Earth the knowledge they had gathered in alien solar systems. This would indeed be a long-term research project, on a scale we can hardly imagine today. It would be as if President Washington embarked upon a scheme which could only benefit Lincoln—or Kennedy. One cannot easily imagine today's Congress voting vast sums for purely scientific projects, whose success or failure could not possibly be known before the year 2200. . . .

So perhaps it is more appropriate, at this moment of time, not to wonder when we will be attempting such feats, but to ask ourselves if other civilizations may not already have achieved them, long ago. The concept of intelligent, advanced life throughout the universe, which until recently was seldom found outside the pages of science fiction, has become respectable with almost explosive suddenness. A number of radio astronomers have seriously suggested that we should be on the lookout for visiting spaceprobes which may have been orbiting round the sun, keeping a watch on the planets, for thousands of years. . . .

Such probes, if they exist, would be far more sophisticated than any that we can build today; they would make STAR seem as primitive as a Greek water clock. Because it would take years or decades for any controlling signals to reach them, they would have to be completely autonomous, able to cope with any emergency or unexpected event. In other words, they would have to be intelligent.

If we encountered such a probe, and got into communication with it, we might not be able to decide whether we were dealing with a form of life—or with a machine. And at this level, indeed, the distinction might turn out to be altogether unimportant, if not meaningless.

All these ideas are likely to remain pure speculation for centuries to come; yet before the no-longer-distant year 2000—in fact, before 1990—our own spaceprobes will have left the Solar System. With any luck, they should continue to broadcast back information for years; just how far we can track them depends on the size of the ground-based radio telescope used to receive the signals. The 210-foot dish of the Deep Space Network should be able to follow the Pioneers out to several times the distance of Pluto. But if we used the great 1,000-foot dish of the Arecibo Observatory in Puerto Rico, the range would be increased fivefold—perhaps to fifty or a hundred *billion* miles. Unfortunately, such huge and expensive instruments will always be busy on so many programs that they cannot easily be diverted to other projects.

We do not know what we may find, in the regions between the stars. Yet every time we have sent ourselves, or our instruments, into new territory, we have made unexpected and important discoveries. Before the space age dawned, astronomers were quite sure that there was nothing between the planets, except occasional meteors. Today, we look upon interplanetary space as a region of great activity—swept by the tenuous million-mile-an-hour gales of the solar wind, pervaded by magnetic fields, lashed by invisible radiations. . . .

The late C. S. Lewis, with whom I had a long and friendly disagreement about the desirability of man's escaping from his world, was much more accurate than the scientists of his time when he wrote in *Out of the Silent Planet,* back in 1938: "He had read of 'Space'; at the back of his thinking for years had lurked the dismal fancy of the black, cold vacuity, the utter deadness, which was supposed to separate the worlds. . . . Now the very name 'Space' seemed a blasphemous libel for this empyrean ocean of radiance in which they swam. He could not call it 'dead,' he felt life pouring into him from it every moment."

As they leave the outskirts of the Solar System, our first short-range interstellar probes will observe the diminution in the Sun's influence—the fading of its magnetic and gravitational fields. At the same time, they will be able to measure the phenomena of the Greater Universe, which until then will have been masked by the effects of the Sun and the planets. They will be able to take the pulse of the Galaxy itself, detect the onrushing debris of exploding novas, look for the dark, dead stars which, some think, may be even more numerous than those that shine.

Some time in the twenty-first century, at some indefinite distance from the Sun, we will lose contact as the power of their signals weakens—though occasionally, when some giant new telescope is brought into operation, they may be momentarily reacquired. But one day their transmitters will fail, and they will be lost forever.

Or perhaps not; there are two other possibilities. As our space-faring powers develop, we may overtake them with the vehicles of a later age and bring them back to our museums, as relics of the early days before men ventured beyond Mars. And if *we* do not find them, others may.

We should therefore build them well, for one day they may be the only evidence that the human race ever existed. All the works of man on his own world are ephemeral, seen from the viewpoint of geological time. The winds and rains which have destroyed mountains will make short work of the Pyramids, those recent experiments in immortality. The most enduring monuments we have yet created stand on the Moon, or circle the Sun; but even these will not last forever.

For when the Sun dies, it will not end with a whimper. In its final paroxysm, it will melt the inner planets to slag, and set the frozen outer giants erupting in geysers wider than the continents of Earth. Nothing will be left, on or even near the world where he was born, of man and his works.

But hundreds—thousands—of light-years outward from Earth, some of the most exquisite masterpieces of his hand and brain will still be drifting down the corridors of stars. The energies that powered them will have been dead for eons, and no trace will remain of the patterns of logic that once pulsed through the crystal labyrinths of their minds.

Yet they will still be recognizable, while the Universe endures, as the work of beings who wondered about it long ago, and sought to fathom its secrets.

PICTURE CREDITS

Columbia House, a Division of Columbia Broadcasting System, Inc.: *reproduction of Figure 22.*

Hale Observatories: *Figure 21.*

Jet Propulsion Laboratory, National Aeronautics and Space Administration: *Figures 3, 4, 5, 6, 10, 11, 12, 15, 16, and 17.*

National Aeronautics and Space Administration: *Figures 7, 8, and 9.*

In the late 1970's, a remarkable event will take place. The outer planets of our solar system — Jupiter, Saturn, Uranus, Neptune, and Pluto—will arrive in orbital alignments that will make it possible for the first time in contemporary space history to send out a spaceprobe to fly by all of them in one grand sweep, taking pictures and gathering information. This robot spaceprobe would leave Earth in 1976, spending some ten years sending back hundreds of computer-reassembled photographs and thousands of miscellaneous data before plunging into intergalactic space. It would tell us more than we have ever known before about the outer solar system and the very nature of space itself.

This unusual book was inspired by the singular possibilities of that probe. A projection of what that robot spacecraft might see as it voyages to the outer planets, BEYOND JUPITER contains the truly extraordinary illustrations of Chesley Bonestell, an artist internationally known for his ability to paint astronomical subjects with such perfection that his work is sometimes mistaken for photographs—even by astronomers. Bonestell has furnished twenty-six paintings, of which fifteen are reproduced in full color. In addition, there are six diagrams and eleven photographs. The paintings are at once beautiful and uncannily accurate in the manner in which they visualize the planets of our solar system both as we know them today—and as they could be revealed by a future probe.

The text is by Arthur C. Clarke, famous for his books and lectures on science and science fiction. Clarke explores what we already know about the outer planets and what a probe might discover there. He examines the technology of the proposed flight and traces the history of the robot spacecraft that have already preceded it, from Luna III to Mariner 9. And he explains why the flight must be taken:

"It has often been said that when we explore space, we are also exploring time. Only when we have flown missions to every part of the Solar System will we have the vital statistics of all its components. And

only then will we be able to turn back the pages of the book of cosmology—to the origins of our own world, and perhaps of the universe itself."

BEYOND JUPITER: an exciting preview of —the worlds of tomorrow.

Chesley Bonestell is an architect, astronomer and artist. A native of California, he worked with numerous distinguished architects in San Francisco and New York before going to England as a special artist on the *Illustrated London News,* **after which he returned to the United States, where he was associated with, among other projects, the building of the Golden Gate Bridge. After World War II, he teamed with author Willy Ley and Wernher von Braun to illustrate books about space, the result being books such as** *The Conquest of Space, Conquest of the Moon, The World We Live In, The Exploration of Mars, Beyond the Solar System,* **and** *Mars* **(with Robert Richardson). Mr. Bonestell has also worked in motion pictures as an artist and an adviser on space exploration.**

Arthur C. Clarke is the author of over forty books of science and science fiction, among which are such titles as *Childhood's End, 2001: A Space Odyssey, The City and the Stars, The Wind from the Sun, The Promise of Space, The Treasure of the Great Reef,* **and** *Report on Planet Three.* **He has received many honors, including his nomination for an Oscar (with Stanley Kubrick) in 1969 for his work on the movie version of** *2001,* **and he is widely recognized as one of the most proficient interpreters of science to the layman, both in print and on the lecture platform. Although a compulsive traveler, he maintains a home in Colombo, Ceylon.**

180°
VENUS
MERCURY
270°
90°
EARTH
0°
SCALE IN LIGHT MINUTES
0
1
2
3
4
5
6
7
8
9